Detailed Review on Recent Advancements in Fused Deposition Modeling (FDM)

I. John Solomon

Dr.P. Sevvel

Published by

BONFRING®

Intellectual Integrity

Detailed Review on Recent Advancements in Fused Deposition Modeling (FDM)

9 788195 381371 >

ISBN 978-81-953813-7-1

Authors

I. John Solomon

Dr.P. Sevvel

Bonfring

309, 5th Street Extension, Gandhipuram,

Coimbatore-641 012,

Tamilnadu, India.

E-mail: info@bonfring.org

Website: www.bonfring.org

Dedicated to the Entire Scientific Research Community

and

To All Research Scholars

Message from Authors & Sincere Thanks

We kindly bring it to the Knowledge of the Readers that, this book entitled "Detailed Review on Recent Advancements & Investigations in Fused Deposition Modeling (FDM)" is a compilation of the various experimental research works carried out in various parts of the World by several researchers in the field of Fused Deposition Modeling (FDM). This book was compiled & framed by the Authors to provide a detailed knowledge about the various recent advancements in the field of FDM. The Book contains a detailed review about the various advancements and experimental investigations carried out in the field of FDM. The Authors at any point of time, anywhere, does not claim that all these experimental works were carried out by them. The Authors of this book expresses their very Sincere Gratitude & Greater Thanks to All the experimental researchers whose experimental works and investigational results are shared and mentioned in this book. Authors also register that, this Book was not compiled with the objective of earning Money or Profit or Fame in any manner, but purely, it was compiled with the objective of providing an awareness to the research community (especially to research scholars) about the field of Fused Deposition Modeling (FDM). The Authors does not have any intention to misuse the experimental research findings of the researches (whose experimental results have been mentioned in this book) and the Authors also very kindly inform that, the Name of each& every Researchers (whose experimental work have been described in this book) along with their experimental research works have been mentioned very clearly in the references section present at the end the book.

Authors Details

I. John Solomon pursuing his doctoral degree in Mechanical Engineering, Anna University and successfully completed his master's degree with first class in Manufacturing Engineering from Anna University, Chennai, India, in 2010. He has a total of 10 years of teaching experience. He is currently working as Assistant Professor, Department of Mechanical Engineering in Panimalar Engineering College, Chennai. He has published 2 research review papers in Scopus indexed International Journals.

Dr.P. Sevvel obtained his Ph.D. in Mechanical Engineering from Anna University, Chennai in 2016 and successfully completed his M.Tech. degree with First Class Distinction in Industrial Engineering from the National Institute of Technology (NIT) Trichy, India, in 2005. He has a total of more than 15 years of teaching experience. He is currently working as Professor, Department of Mechanical Engineering College in S.A. Engineering College, Chennai. He has a total of 3 ongoing funded Projects worth Rs. 46,90,107 under various schemes of DST, TNSCST He has 4 Indian PATENT GRANTS in his name and has published more than 9 Patents for various Innovative Products including Desktop 3D Printer, A4 Laser Engraver, Multi Terrain Quadcopter, IOT controlled drainage channel cleaning device etc. He is a Life Member of the Indian Society for Technical Education (ISTE) and a Corporate Member in The Institution of Engineers (IE), India. He has published more than 46 research in various SCI & Scopus indexed International Journals. He is currently guiding a total of 9 Research Scholars pursuing their Ph.D. in Centre for Research, Anna University. He is a reviewer for various reputed Journals including International Journal *Materials & Design, Journal of Mechanical Science and Technology*. He has many awards including ACADEMIC PROFESSIONAL ICON AWARD from Institution of Engineers (IE) India, TVLC during the Grant Event TISLA Awards 2019, BEST RESEARCHER AWARD for my distinct contribution in the field of Materials Science during the 6th Annual Millennium Impact Award 2020. He has successfully organized several International Conferences including DST – SERB sponsored ICRDME 2019, ICMEET 2015 etc.

TABLE OF CONTENTS

CHAPTER NO	TITLE	PAGE NO
1	**INTRODUCTION**	**01**
	1.1. Introduction	01
	1.2. Brief Insight into Additive Manufacturing	02
	1.3. History of Additive Manufacturing Process	03
	1.4. Comparison between Additive and Conventional Manufacturing Process	05
	1.4.1. Creating Complex Geometries	05
	1.4.2. Material Wastage	06
	1.4.3. Customization of Product	06
	1.4.4. Difference in Equipment Required	06
	1.4.5. Differences in Business Model	07
	1.4.6. Different Manufacturing Model	07
	1.5. Advantages and Drawbacks of Additive Manufacturing Process	08
	1.5.1. Advantages of Additive Manufacturing	09
	1.5.2. Disadvantages of Additive Manufacturing	11
	1.6. Applications of Additive Manufacturing	12
2	**APPLICATIONS OF ADDITIVE MANUFACTURING**	**16**
	2.1. Classification of Additive Manufacturing	16
	2.2. VAT Photo Polymerization	17
	2.2.1. Working Principle	17
	2.2.2. Materials Used	18
	2.2.3. Applications	18

	2.3. Material Extrusion	19
	2.3.1. Working Principle	19
	2.3.2. Materials Used	21
	2.3.3. Applications	21
	2.4. Powder Bed Fusion	21
	2.4.1. Working Principle	21
	2.4.2. Materials Used	23
	2.4.3. Applications	23
	2.5. Material Jetting	24
	2.5.1. Working Principle	24
	2.5.2. Types of Material Jetting	25
	2.5.3. Material Used	27
	2.5.4. Applications	27
	2.6. Binder Jetting	28
	2.6.1. Working Principle	28
	2.6.2. Post Processing Steps in Binder Jetting Process	29
	2.6.3. Material Used	30
	2.6.4. Applications	30
	2.7. Direct Energy Deposition	31
	2.7.1. Working Principle	31
	2.7.2. Materials Used	32
	2.7.3. Applications	33
	2.8. Sheet Lamination	33
	2.8.1. Working Principle	33
	2.8.2. Materials Used	34
	2.8.3. Applications	34

3	**FUSED DEPOSITION MODELLING**	**35**
	3.1. Introduction to Fused Deposition Modeling [FDM]	35
	3.2. Working Principle of Fused Deposition Modeling	36
	3.3. Parts of a FDM 3D Printer	37
	3.3.1. Extruder	37
	3.3.2. Types of Extruder	41
	3.3.3. Structure	42
	3.3.4. Printing Surface	47
	3.3.5. Electronics	48
4	**MATERIALS USED IN FUSED DEPOSITION MODELING**	**50**
	4.1. Introduction to Filament Materials	50
	4.2. Properties of Polymers for FDM	50
	4.3. Materials Used in FDM	52
	4.3.1. Polylactic Acid (PLA)	52
	4.3.2. Acrylonitrile Butadiene Styrene (ABS)	53
	4.3.3. Polycarbonates (PC)	55
	4.3.4. Polyethylene Terephthalate (PET)	57
	4.3.5. PETG	59
	4.3.6. Thermoplastic Polyurethane (TPU)	61
	4.3.7. Acrylonitrile Styrene Acrylate (ASA)	63
	4.3.8. ULTEM	64
5	**RESEARCH TRENDS IN FDM OF THERMOPLASTIC POLYMERS - ABS**	**67**
	5.1. Introduction to Thermoplastic Polymers	67
	5.2. Acrylonitrile Butadiene Styrene Polymer in FDM	68

	5.2.1. Raster Width	68
	5.2.2. Air Gap	69
	5.2.3. Model Build Temperature	69
	5.2.4. Raster Orientation	70
	5.3. FDM Printed ABS Parts	70
	5.4. Injection Molds Vs FDM Printed ABS Part	72
	5.4.1. Tensile Test	72
	5.4.2. Shear Test	74
	5.4.3. Compression Test	76
	5.4.4. Flexural Test	77
	5.4.5. Thermo Gravimetric Analysis (TGA)	79
6	**RESEARCH TRENDS IN FDM OF THERMOPLASTIC POLYMER - PLA**	**80**
	6.1. Introduction	80
	6.2. 3D Printing of PLA Composites through Fused Deposition Modeling	81
	6.3. Other Directions in 3D Printing of PLA based Materials	82
	6.4. Influence of Polymer Formulation on the Printing Quality and Mechanical Properties	84
	6.4.1. Rheological Properties of PLA Filaments	85
	6.5. Surface Quality in FDM Process	86
	6.5.1. Experimental Plan	89
	6.5.2. Surface Roughness Evaluation	90
	6.5.3. Experimental Findings	90

	6.6. Tensile Properties of PLA Specimen	91
	6.6.1. Material and Manufacturing Process	91
	6.6.2. Tensile Test	92
	6.6.3. Experimental Work Findings	93
	6.7. Compression Test of PLA Specimen	93
	6.7.1. Specimens for Compression Test	94
	6.7.2. Results for Compression Test	95
7	**RESEARCH TRENDS IN ULTEM POLYMER**	**98**
	7.1. Introduction	98
	7.2. Materials	98
	7.3. Density	99
	7.4. Tensile Test	100
	7.5. Effect of Build Parameters	103
	7.5.1. Specimen Details	103
	7.5.2. Design of Experiments – Factors and Levels	103
	7.5.3. Effect of Build Style and Temperature	104
	7.5.4. Effects of Build Direction, Raster Angle and Air Gap	106
	7.5.5. Effect of Build Direction and Raster Angle	107
	7.5.6. Effect of Air Gap	109
	7.6. Compression Test	110
	7.7. Flexural Behavior	112
8	**RESEARCH TRENDS IN POLYMER COMPOSITES**	**116**
	8.1. Introduction	116
	8.2. FDM of Thermoplastic Composites	117

	8.3. FDM of ABS Composites	118
	8.3.1. Carbon based Reinforcement	118
	8.3.2. Graphene based Reinforcement	120
	8.4. FDM of PLA Composites	121
	8.5. FDM of Polyamide	125
	8.6. FDM of Polypropylene	127
	8.7. FDM of Polyethylene	129
	8.8. FDM of Polyether Ether Ketone (PEEK)	131
9	**RECENT TRENDS IN FUSED DEPOSITION MODELING OF METALS**	**132**
	9.1. Introduction	132
	9.2. Disadvantages of Metal Additive Manufacturing	132
	9.3. Metal Extrusion Process	132
	9.4. Key Considerations	134
	9.4.1. Existing Designs May Not be Directly Transferable	134
	9.4.2. Geometric Complexity is as Asset	134
	9.4.3. Start by Defining the Basic Requirements	134
	9.4.4. Always Use Minimum Support Structures	134
	9.4.5. Post-processing is Always Required	134
	9.5. Creating Metal Parts by Fused Deposition Modeling and Sintering	135
	9.5.1. Experimental Methods	136
	9.5.2. Microstructure Analysis	136
	9.5.3. Hardness and Tensile Strength	137
	9.6. Fabrication of Iron - Nylon Composite Filament	138

	9.6.1. Tensile Properties	140
	9.7. Metal-Polymer Composite Filament	142
	9.7.1. Effects of Viscosity on Tensile Properties	143
	9.7.2. Effects of Fill Density on Tensile Properties	145
	9.7.3. Effect of Metal Particle Infill on Tensile Properties	145
	9.7.4. Effect of Metal Particle Infill on Thermal Conductivity	147
	9.8. Polymer Metal Composite (PE/Cu)	148
	9.8.1. Flexural Strength Test	151
10	**FUTURE RESEARCH SCOPE AND CONCLUSION**	**154**
	10.1. Future Research Scope	154
	10.2. Conclusion	159
	REFERENCES	**160**

LIST OF FIGURES

S. NO	DESCRIPTION	PAGE NO
1.1	Component Manufactured by Additive Manufacturing Technique	01
1.2	Steps in Part Creation in Additive Manufacturing Process	02
1.3	Metal Bracket Manufactured by Additive Manufacturing Process	03
1.4	Graphical Representation of François Willème (1830-1905)	03
1.5	Complex Automotive Component Manufactured by AM Process	05
1.6	Product Customization	06
1.7	A 3D Printing Equipment	07
1.8	Example for Complex Design	09
1.9	A Prototype Model	09
1.10	3D Printed Bio Part	11
1.11	3D Printed Prosthetic ARM	13
1.12	Example for Components Manufactured by AM Technique	14
1.13	Additive Fuel Nozzle	14
1.14	Additive Manufacturing in Tooling	15
1.15	Expansion of The Lewis Grand Hotel in the Philippines	15
2.1	Types of Additive Manufacturing Processes	16
2.2	An Example for VAT Photo Polymerization Process	17
2.3	Types of Processes in VAT Photo Polymerization	18
2.4	Component Manufactured by Stereolithography Process	19
2.5	Schematic Representation of Material Extrusion Process in AM	20
2.6	Different Types of Material Extrusion based AM Process	20
2.7	Powder Bed Fusion Process	22

2.8	Different Variants in Powder Bed Fusion Process	22
2.9	Cranial Implant Manufactured by PBF Process	23
2.10	Schematic Representation of Material Jetting Process	24
2.11	Types of Material Jetting Process	25
2.12	Schematic Representation of Poly Jet Printing Process	26
2.13	Ceramic Components Manufactured by Nano Particle Jetting	26
2.14	Drop on Demand Material Jetting Process	27
2.15	Schematic Representation of Binder Jetting Process	28
2.16	The M5 Screws 3D printed Using Binder Jetting Process	29
2.17	Post Processing Steps in Binder Jetting Processes	30
2.18	Example of Components Manufactured by Binder Jetting Process	31
2.19	Schematic Representation of Direct Energy Deposition Process	32
2.20	Illustrates How DED Method is Used for Repair of Gear Teeth	33
2.21	Schematic Representation of Sheet Lamination Process	34
3.1	Illustrative Diagram of FDM	35
3.2	Illustrates the Market Size of Various 3D Printing Processes	36
3.3	Schematic Representation of FDM Process	37
3.4	FDM Extruder Assembly Setup	38
3.5	Extruder Hotend	38
3.6	Angled Fan Mount on Extruder Hotend	39
3.7	Thermal Bar Used in FDM	39
3.8	Heating Block Made of Aluminium	40
3.9	Brass Nozzles Used in FDM Processes	40
3.10	Schematic Representation of Gear Type Extruder	41
3.11	Schematic Representation of Direct and Bowden Extrusion	42

3.12	Schematic Illustration of Cartesian Type 3D Printer	43
3.13	3D Delta Metal Printer	44
3.14	Commercial Polar Type FDM 3D Printer	45
3.15	Robotic ARM 3D Printer	46
3.16	Shows a Component being Printed on a Glass Print Bed	47
3.17	Simple 3D Printer Setup with Various Electronic Components	49
4.1	Filament Materials Used for Printing	50
4.2	Decision Criteria for Material Selection	51
4.3	A 3D Printing Filament	52
4.4	3D Printed PLA Sample	53
4.5	Process of Laying ABS Layer by Layer	54
4.6	Component Printed from ABS Filament in a FDM Process	55
4.7	Custom Assembly Pallet Made from PC	56
4.8	Lid Made from PET on FDM Mold	58
4.9	PETG Filament and 3D Printed Vacuum Nozzle	59
4.10	3D Print with PETG	60
4.11	Thermoplastic Polyurethane Part Manufactured with FDM	61
4.12	UV Resistant Component Printed from ASA Thermoplastic	63
4.13	Ultem 9085 Filament Used for 3D Printing in FDM	65
4.14	Automotive Component Made from Ultem Filament	66
5.1	Simplified Process flow Chart of FDM Including different Types of Software's Used	68
5.2	Illustrating Raster Width, Air Gap and Raster Angle	69
5.3	Schematic Representation of Various Types of Air Gaps	69
5.4	Various Raster Orientations in Fused Deposition Modeling	70

5.5	Surface Finish of a FDM Printed ABS Component	71
5.6	Tensile Strength of FDM with Injection Moulded ABS Part with Zero Air Gap	72
5.7	Tensile Strength of FDM with Injection Moulded ABS Part with Negative Air Gap	73
5.8	Failure Mode of Specimens with Various Raster Orientations	73
5.9	Scanning Electron Microscope Pictures of Fractured Surface with Negative Air Gap	74
5.10	Average Stress-strain Behavior of Ten [+45/-45] Flat Orientation ABS Samples	76
5.11	Compression Specimens with Two different Build Directions	76
5.12	Compressive Strength of Specimens with different Build Direction Compared with Injection Molded ABS	77
5.13	3-point Flexural Loading System	78
6.1	Single PLA Polymer Monomer Unit	80
6.2	PLA/HA Nano Composites by FDM 3D Printer	81
6.3	Schematical Representation Showing the Yoda Sample While Printing	84
6.4	Overhang Printing Defects Due to Uncontrolled Flow of Polymer	85
6.5	(a) Shear (Circles) and Complex (Triangles) Viscosity for White (Open Symbols) and Green (Full Symbols) polylactide (PLA). (b) Elastic (Circles) and Viscous (Triangles) Modulus for White (Open Symbols) and Green (Full Symbols) PLA	86
6.6	WITBOX Printer and PLA Filament	87
6.7	Surface Roughness Measuring Setup	88
6.8	Printed Samples for Experiment	88
6.9	Details of Measurement Procedure	90
6.10	Geometrical Parameters of DB Specimen	92

6.11	Printing Orientation of 3D Specimen	92
6.12	Tensile Specimen Before (a) and After (b) Tensile Test	93
6.13	Geometrical Data (In MM) of the Parallelepiped-shaped Specimen for the Compression Tests	94
6.14	Printing Plane for Parallelepiped Specimens for Compression Tests	95
6.15	Stress-strain Compression Test Curves for the Test Specimens	96
7.1	FDM Samples Built Direction	99
7.2	Macroscopic Views of Samples (a–c) and Fracture Surfaces (d–f)	102
7.3	Comparison of Compression Properties for Solid and Sparse Specimens	105
7.4	Yield Strength of Horizontal Built and Vertical Built Solid Specimens	105
7.5	Interaction of Build Direction and Raster Angle on Yield Strength of Solid Specimens	107
7.6	Sample Stress-strain Curves for Horizontal and Vertical Build Specimens	107
7.7	Physical Model Horizontal Build Solid Model with (0°, 90°) and (45°, -45°) Raster Angles	108
7.8	Physical Model Vertical Build Solid Model with (0°, 90°) and (45°, -45°) Raster Angles	108
7.9	Effect of Air Gap on Yield Strength for Solid Specimens	109
7.10	Stress-strain Curves at different Air Gaps for Horizontal-build and Vertical-build Specimens, Where the Raster Angle was (0°, 90°)	109
7.11	Graph Of Average Stress-Strain Results for Compression Test	110
7.12	Graph of Average Stress-strain Results for Flexural Test	112
7.13	Light Microscope Image Showing Dense Sample Built in Flat Direction With 0° Layer Orientation	113

7.14	Light Microscope Image Showing Samples with Flaws when Built in Flat Direction with +45°/- 45° Layer Orientation	114
7.15	Light Microscope Image of Incomplete Fills for the Sample with the 90° Layer Orientation Printed in the Edge Direction	114
8.1	Schematic of FDM Process Using Continuous Fiber Reinforcement a) In-Nozzle Impregnation b) Printing with Two Separate Nozzle	117
8.2a	The Presence of Voids in the FDM Printed Composite Parts a) ABS+10 wt. % Short Carbon Fibers of 150 μm Length	119
8.2b	The Presence of Voids in the FDM Printed Composite Parts b) Neat ABS c) ABS+10 wt % Short Carbon Fibers of 0.35 mm Length	119
8.3	Microstructure of 3D Printed (a) PLA and (b) PLA-graphene Parts. Fractured Cross-sectional. Microstructure of 3D Printed (c) PLA and (d) PLA-graphene	122
8.4	The Interface is Locally Heated to Allow for Polymer Diffusion and Increased Fracture Strength	123
8.5	a) Thermoplastic PLA Filament Coated with CNT-rich Layer b) After the Microwave Heating, The Interface is Locally Heated to Allow for Polymer Diffusion and Increased Fracture Strength	124
8.6	Illustration of the Failure Mechanism of 3D Printed Continuous Fiber Reinforced Composites, Tensile Behavior: a) and b), Flexure Behavior of Carbon FRTP: c) and d), Flexure Behavior of Glass FRTP: e) and f)	127
8.7	FDM Printed Polypropylene a) with Distortion and Decoupling from the bed, b) Reduced Distortion and Improved Adhesion to the Bed at Optimal Printing Parameters	128
8.8	a) Shrinkage of the FDM Printed Polypropylene Part b) Reduction in Shrinkage of PP with 30 wt%. Natural Fiber Hemp	129

8.9	SEM Images of the Surface (left) and Cross Sectional Area (Right) of HDPE Specimens Obtained by FFF (1–4) and by Injection Molding, Respectively (5–6). In Contrast to Regular Printing (1 and 2) FFF with Gradually Adjusted Filling Amount (3–4) is Clearly Superior in Terms Surface Quality and Preventing Defect Formation	130
9.1	Metal Parts Manufactured by Fused Deposition Modeling	133
9.2	Schematic Illustration of FDMS	135
9.3	SEM Images of the Cross-sections of a) Filament and b) Green Part	136
9.4	Optical Microscopy of Microstructure in FDMS 316L Part	137
9.5	Filament Produced by Iron – Nylon Composite	139
9.6	Comparison of Stress Strain Curves for Iron Nylon Composites	141
9.7	Examples of Parts and Inserts Produced on the FDM System	142
9.8	DSC Analysis Result of ABS	143
9.9	Viscosity (μ) of ABS at different Temperatures	144
9.10	Tensile Properties of the 3D Structures Manufactured by ABS at Different Temperatures	144
9.11	Tensile Strength of ABS Specimens with Increasing Fill Density	145
9.12	Stress–strain Curves for ABS with Copper Particles	146
9.13	Stress–strain Curves for ABS with Iron Particles	146
9.14	Microscopic Structure of Cutting Plane of Specimens: (a) ABS-Cu10wt% (b) ABS-Cu30wt%	147
9.15	Coefficient of Thermal Expansion by Changing the Copper Content	147
9.16	Thermal Conductivity of ABS-Cu Filaments as a Function of Copper Content	148
9.17	Filament Production Line	149
9.18	Schematic Illustration of Filament Production Dies	149

9.19	(a, b) Distortion of PE-based Specimens	150
9.20	Flexural Strength Test	152
9.21	Flexural Strength Test Results	152
9.22	Schematic Representation of Bending Test	153

List of Tables

S. NO	DESCRIPTION	PAGE NO
1.1	Historical Developments in Additive Manufacturing Process	04
1.2	Difference Between Additive and Traditional Process	08
5.1	Shear Modulus, Yield Strength, and Ultimate Strength Average Values with 95% CIs for the ABS	75
5.2	Flexural Test Results	79
7.1	Average Value of Densities and Porosity of Printed Parts for the Different Build Orientation	99
7.2	ANOVA Test for Tensile Properties	100
7.3	Results of Tensile Properties of Ultem9085 Samples Built According the Three Building Directions	101
7.4	Factors and Levels for Solid Build Style	104
7.5	Factors and Levels for Sparse Build Style	104
7.6	Effects Table from JMP 11 for Solid Specimen	106
7.7	Average Results of Compression Extension, Compressive Strength, Compressive Strain and E-modulus	111
7.8	Average Results for Three-point Bending Load	112
9.1	Tensile Modulus, Tensile Stress at Peak and Tensile Strain at Break	140
9.2	The Volumetric Flow Rate and Viscosity of ABS at Various Temperatures	143
9.3	Tensile Properties of the Specimen	146
9.4	Optimum Parameter for Filament Production	149
9.5	Optimum Condition for Printing	151
10.1	Road Map for Future Research	158

CHAPTER 1

1. Introduction

1.1. Introduction

Additive manufacturing is a relatively recent manufacturing method which has become a key area of interest in multiple industrial sectors. Deriving from CAD models the process can be used to create solid yet highly complex parts and pushes towards a tool-less manufacturing environment meaning improved quality and better efficiency in many cases [1].

Additive manufacturing (AM) is the process of joining materials to make objects from Computer Aided Design (CAD) model data, usually layer upon layer, as opposed to subtractive manufacturing methods. Additive manufacturing is also called 3D printing, additive fabrication, or freeform fabrication.

Additive Manufacturing can bring digital flexibility and efficiency to manufacturing operations [2]. Additive manufacturing is a novel method of manufacturing parts directly from digital model by using layer by layer material build-up approach. This tool-less manufacturing method can produce fully dense metallic parts in short time, with high precision. Features of additive manufacturing like freedom of part design, part complexity, light weighting, part consolidation and design for function are garnering particular interests in metal additive manufacturing for aerospace, oil & gas, marine and automobile applications. The figure 1.1 given below illustrates an example of component being manufactured by one of the additive manufacturing process.

Additive manufacturing uses data computer-aided-design (CAD) software or 3D object scanners to direct hardware to deposit material, layer upon layer, in precise geometric shapes. As its name implies, additive manufacturing adds material to create an object.

Fig. 1.1: Component Manufactured by Additive Manufacturing Technique [3]

By contrast, when you create an object by traditional means, it is often necessary to remove material through milling, machining, carving, shaping or other means. Additive manufacturing sometimes referred to as rapid prototyping or 3D printing, are technologies that use successive layers of material to create 3D objects directly from a computer generated model. Although Additive manufacturing process flow vary between the 7 different additive manufacturing technologies it uses to create the 3D parts, each broadly follows these common steps given in figure 1.2 to create the final part. The steps starts with 3D model creation, STL file creation, conversion and transfer to machine till the build process is completed which is followed by a part removal and post processing section which plays an important role in this process.

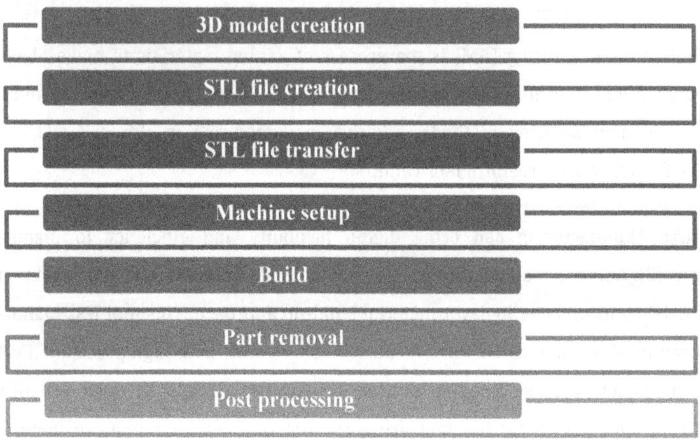

Fig. 1.2: Steps in Part Creation in Additive Manufacturing Process [4]

1.2. Brief Insight Into Additive Manufacturing

A computer-aided design (CAD) is created and exported to stereolithography (STL) file format that is read by the AM equipment. There are many techniques available, which can be categorized according to their raw material. They are: (1) powder-based, (2) liquid-based, and (3) solid based. Some examples of powder-based techniques include selective laser melting (SLM), selective laser sintering (SLS), and electron beam melting (EBM). Liquid-based techniques include stereolithography apparatus (SLA) and polyjet while solid-based techniques include laminated object manufacturing (LOM) and fused deposition modeling (FDM). Basically, any material can be produced by one or another AM technique today. These materials can be divided into four main categories: plastics, metals, ceramics, and composites. The figure 1.3 given below shows a metal bracket manufactured by additive manufacturing process.

Fig. 1.3: Metal Bracket Manufactured by Additive Manufacturing Process [5]

Today, a variety of plastics with vastly different mechanical, chemical and environmental properties can be additively manufactured, Polyimides are a group of polymers with exceptionally high heat and chemical resistance that are yet to be used in AM. These materials vary in transparencies, thermal or mechanical properties. Polyamides are the most popular thermoplastics used in plastic laser sintering due to their widespread use in injection molding. However, the specific grades of polyamides used in AM have different physical properties and wider processing windows as compared to their injection molding counterparts even though they are chemically identical. The range of polymers used in AM encompasses thermoplastics, thermosets, elastomers, hydrogels, functional polymers, polymer blends, composites, and biological systems.

1.3. History of Additive Manufacturing Process

The initial dabbling with Additive Manufacturing ideology dates back as far as the 19th Century; 1859 to be precise, when François Willème (graphical illustration shown in figure 1.4), a French photo sculptor used 24 cameras to simultaneously photograph objects from different angles[6].

Fig. 1.4: Graphical Representation of François Willème (1830-1905)

This was the first demonstration of 3D scanning technology. But the kick-start of AM began in 1981, when Hideo Kodama of the Nagoya Municipal Industrial Research Institute, published information regarding the manufacturing of a solid printed model. In the 1980s, 3D printing techniques were considered suitable only for the production of functional or aesthetic prototypes, and a more appropriate term for it at the time was rapid prototyping. As of 2020, the precision, repeatability, and material range of 3D printing has increased to the point that some 3D printing processes are considered viable as an industrial-production technology, whereby the term additive manufacturing can be used synonymously with 3D printing[7]. One of the key advantages of 3D printing is the ability to produce very complex shapes or geometries that would be otherwise impossible to construct by hand, including hollow parts or parts with internal truss structures to reduce weight. Fused deposition modeling, or FDM, is the most common 3D printing process in use as of 2020. The development history of Additive manufacturing process is illustrated in the table 1.1 given below.

Table 1.1: Historical Developments in Additive Manufacturing Process [6]

Year	Contributor	Development
1859	Francois Willeme,	First demonstration of 3D scanning technology.
1892	Joseph E. Blanther	Files a patent for the creation of 3D topographical maps using a layering method
1980	Dr. Hideo Kodama	Photopolymer rapid prototyping technique
1984	Charles (Chuck)	Filed a patent for Stereolithography
1986	Charles Hull	Becomes co-founder of the world's first 3D printing company- 3D Systems.
1987		3D Systems create the world's first Stereolithographic apparatus
1988	Carl Deckard.	Patents a Selective Laser Sintering (SLS) technology
1989	Scott Crump, Co-Founder of Stratasys Inc.	Filed a patent for Fused Deposition Modeling (FDM).
1992		DTM produced the world's first selective laser sintering (SLS) machine, this shoots a laser at a powder instead of a liquid.
1999		The birth year of medical 3D Bio printing- the first organ (a urinary bladder) was successfully transplanted to a patient.
2006		First SLS Machine becomes commercially viable.
2011		World's first printed aircraft and prototype car are developed.
2014		Carl Deckard's SLS Patent Expires. The low cost, high resolution technology brings new opportunities for 3D printing companies to expand.
2015	Cellink, a Swedish company	Launches the first commercial bio-ink which can be used for printing tissue cartilage.
2018		3D printing in construction gains momentum as houses are being printed to solve the housing crisis globally. Architecture is also reimagined with 3D printing potential.

1.4. Comparison between Additive and Conventional Manufacturing Process

What makes additive manufacturing process a unique one?
Additive manufacturing operates by adding layers of material together to make an object. Traditional manufacturing methods, by contrast, are subtractive in nature. Subtractive manufacturing involves removing parts of a block of material in order to create the desired shape.

1.4.1 Creating Complex Geometries

The most important of aspect of additive manufacturing is that it can be used to create complex geometries that would be difficult or impossible to achieve with traditional manufacturing methods. These complex geometries are often stronger and lighter than their traditionally manufactured counterparts. The design freedom offered by AM has many advantages over its counterpart [8]. Strongly related to its ability to create complex geometries more easily than other manufacturing methods is the fact that additive manufacturing eliminates the additional costs normally associated with creating more complex objects. With traditional methods, a highly complex part typically costs much more to make than a very simple one. But in additive manufacturing the process is identical regardless of the complexity of a part. Thus, additional cost is no longer a factor. Figure 1.5 given below shows a complex automotive component manufactured by AM technique.

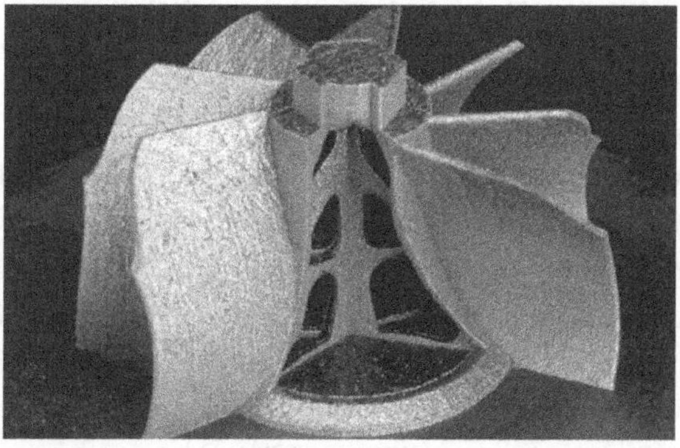

Fig. 1.5: Complex Automotive Component Manufactured by AM Process [9]

1.4.2 *Material Wastage*

Another major functional difference between additive and traditional manufacturing methods is the fact that additive manufacturing involves far less material waste. We effectively render a material waste byproduct when removing it in a subtractive manufacturing process. Additive manufacturing only uses the material necessary to create a part, ensuring that waste is very minimal. Parts that require support structures for printing, however, will still involve an appreciable amount of wasted material [10].

1.4.3 *Customization of Product*

A major functional difference between additive manufacturing and other manufacturing methods is the fact that additive manufacturing can be used to easily create customized products. Mass manufacturing techniques that companies currently employ create large numbers of identical products but offer little room for customization. In contrast, given their ability to print objects of any configuration from scratch, 3D printers are ideal for mass customization [11]. Figure 1.6 illustrates an example for customization.

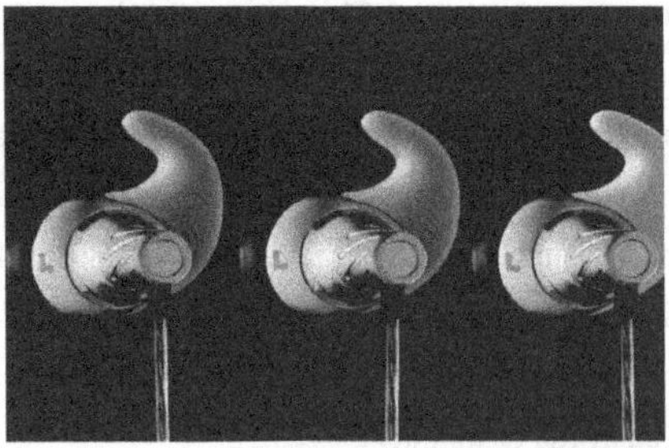

Fig. 1.6: Product Customization [12]

1.4.4 *Difference in Equipment Required*

Additive manufacturing also differs from traditional manufacturing methods in terms of the technology that we use to facilitate it. Traditional manufacturing processes can often require several steps, each using a different machine. In metal machining, for example, turning, milling and drilling are frequently used in concert to create a finished metal part. In additive manufacturing, however, a single device handles all aspects of creation. By extension, additive manufacturing involves less labor than traditional manufacturing. 3D printers create parts in a completely automated manner, requiring little oversight from an operator. By contrast, traditional manufacturing requires a different skilled laborer to operate each individual machine used. Figure 1.7 shows an real time 3D printer.

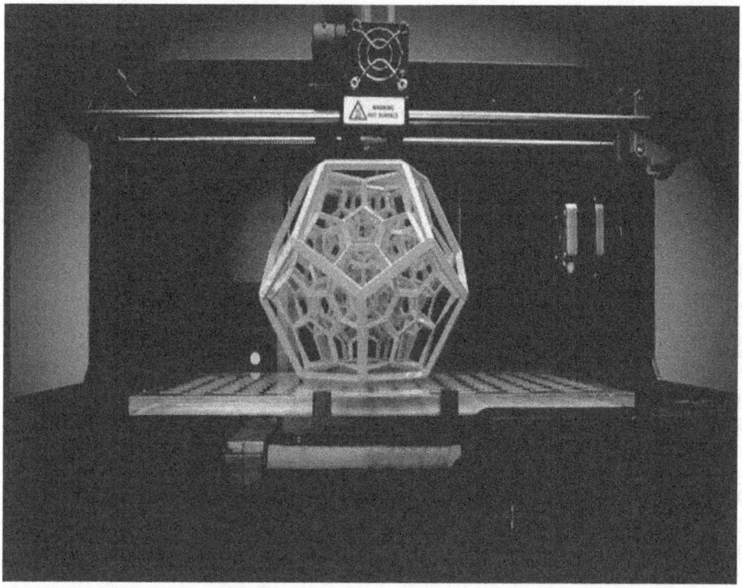

Fig. 1.7: A 3D Printing Equipment [13]

1.4.5 Differences in Business Model

Additive manufacturing lowers the barrier to entrepreneurship and product development. Using 3D printers, businesses can prototype and produce their products using a single piece of equipment, rather than investing in an entire factory setup. In fact, thanks to a host of 3D printing service companies that have arisen in the past several years, small businesses can even make use of additive manufacturing without the initial cost of buying the equipment themselves.

1.4.6 Different Manufacturing Model

The greatest difference between additive and traditional manufacturing is decentralization. We can deploy the 3D printing devices that make additive manufacturing possible locally, rather than in large, centralized factories. As a result, manufacturing can move closer to end users. Through this model, manufacturers can shorten their supply chains and cut down on the costs associated with shipping logistics.

The major difference between the additive manufacturing process and traditional or conventional manufacturing process is summarized in the table 1.2.

Table 1.2: Difference between Additive and Traditional Process [14, 15]

Additive Manufacturing	**Traditional Manufacturing**
It is a layer by layer material addition process	It is a material removal process
Easy to manufacture complex parts	Difficult to manufacture complex parts
Material wastage is less when support structures are not needed.	Material wastage is more.
Full moving assemblies can be printed.	Difficult to manufacture full moving parts.
Highly customized products can be manufactured.	High customization is not possible.
Involves less number of skilled labors.	Need highly skilled labors.
Products can be manufactured very closer to the end user.	It is not possible to manufacture products closer to end user.
It is a highly decentralized process where products can be manufactured locally.	It is a highly centralized process
Best suited for smaller items and parts especially in plastics.	Best suited for manufacturing voluminous items and parts, especially in metals.
Layering often leaves a slightly stepped or rough surface which needs to be finished post printing by sanding or blowing.	A variety of surface finish can be obtained.
Uses computers and specialist 3D printing, direct digital manufacturing, rapid prototyping or additive and layered fabrication.	Uses computers and robotics to assist standard machining processes.
Shorter production time.	Longer production time.
These processes are applicable to a narrow range of materials.	These processes can efficiently handle a wide range of materials.
Consolidation of complex assemblies into single complex part feasible.	Consolidation of parts not feasible.
Rapid design changes.	Fixed design and slow changeovers.

1.5. Advantages and Drawbacks of Additive Manufacturing Process

Additive Manufacturing, which has grown in leaps and bounds in the past 5 years, is likely to continue as a huge technology trend in the current decade. And the industry believes that that AM is a significant technology in the third industrial revolution. As the costs associated with AM have decreased and its practicality has increased, the technology has captured both the interest and imagination of consumers and scientists alike. Several technology trends are fueling the evolution of AM and are enhancing its abilities and offerings [16]. Since the introduction of 3D printing in the 1980s, the benefits of producing small amounts of complex parts have been well understood to manufacturers. With new technology coming out and so many companies turning toward additive manufacturing in recent years, more and more advantages are being revealed on a regular basis [17].

1.5.1 *Advantages of Additive Manufacturing*

Additive manufacturing process offers a range of advantages compared to traditional manufacturing methods. These advantages include those related to design, time and cost, amongst others.

1. Flexible Design

Additive Manufacturing processes give designers the flexibility to push the limits of engineering design beyond the constraints of traditional manufacturing methods. More traditional processes have design restrictions which no longer apply with the use of Additive Manufacturing. Figure 1.8 illustrate a graphical example for complex design

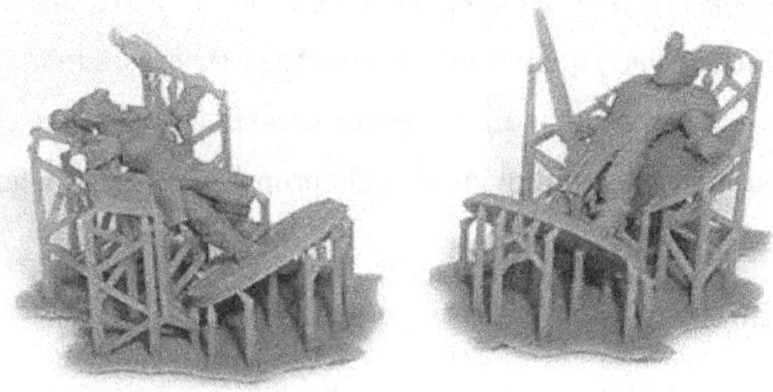

Fig. 1.8: Example for Complex Design [18]

2. Rapid Prototyping

Additive manufacturing machine can manufacture parts within hours, which speeds up the prototyping process. This allows for each stage to complete faster. When compared to machining prototypes, 3D printing is inexpensive and quicker at creating parts as the part can be finished in hours, allowing for each design modification to be completed at a much more efficient rate. Figure 1.9 shows a rapid prototype manufactured by AM.

Fig. 1.9: A Prototype Model [19]

3. Print on Demand

Print on demand is another advantage as it doesn't need a lot of space to stock inventory, unlike traditional manufacturing processes. This saves space and costs as there is no need to print in bulk unless required. The 3D design files are all stored in a virtual library as they are printed using a 3D model as either a CAD or STL file, this means they can be located and printed when needed. Edits to designs can be made at very low costs by editing individual files without wastage of out of date inventory and investing in tools [20].

4. Strong and Light Weight Parts

The main additive manufacturing material used is plastic, although some metals can also be used for printing. However, plastics offer advantages as they are lighter than their metal equivalents. This is particularly important in industries such as automotive and aerospace where light-weighting is an issue and can deliver greater fuel efficiency. Also, parts can be created from tailored materials to provide specific properties such as heat resistance, higher strength or water repellency.

5. Faster Design and Production

Depending on a part's design and complexity, objects can be printed within hours, which is much faster than molded or machined parts. It is not only the manufacture of the part that can offer time savings through additive manufacturing but also the design process can be very quick by creating STL or CAD files ready to be printed [21].

6. Minimizing Waste

The production of parts only requires the materials needed for the part itself, with little or no wastage as compared to alternative methods which are cut from large chunks of non-recyclable materials. Not only does the process save on resources but it also reduces the cost of the materials being used. With AM, manufacturers can print parts on demand. This reduces the need for tooling and reduces the amount of material required to produce components. Solid metals require forming and milling, both processes that require large amounts of energy and create material waste. As AM enables components to be printed on demand, no finished product goes unsold, so there is no over-production [22].

7. Cost Effective

As a single step manufacturing process, Additive manufacturing saves time and therefore costs associated with using different machines for manufacture. Additive manufacturing machines can also be set up and left to get on with the job, meaning that there is no need for operators to be present the entire time. As mentioned above, this manufacturing process can also reduce costs on materials as it only uses the amount of material required for the part itself, with little or no wastage.

8. Advanced Healthcare

3D printing is being used in the medical sector to help save lives by printing organs for the human body such as livers, kidneys and hearts. Further advances and uses are being developed in the healthcare sector providing some of the biggest advances from using the technology. Figure 1.10 shows an AM component used in health care sector.

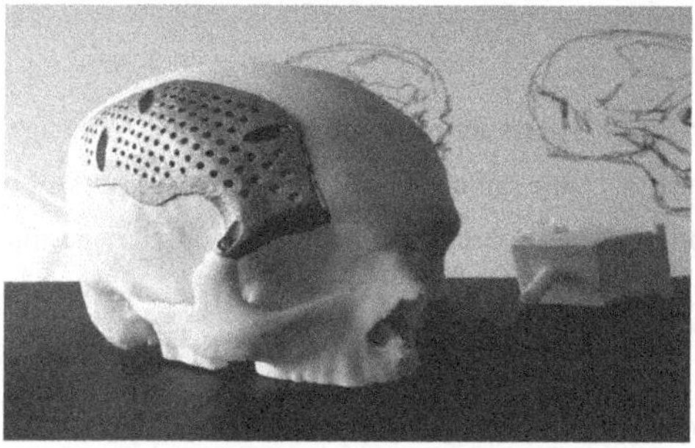

Fig. 1.10: 3D Printed Bio Part [23]

1.5.2 Disadvantages of Additive Manufacturing

While additive manufacturing and 3-D printing as a whole have been largely integrated into today's society and especially the medical field, there are some limitations and disadvantages to the process.

1. Printing Materials

While 3D Printing can create items in a selection of plastics and metals the available selection of raw materials is not exhaustive. This is due to the fact that not all metals or plastics can be temperature controlled enough to allow 3D printing. In addition, many of these printable materials cannot be recycled and very few are food safe.

2. Build Size

3D printers currently have small print chambers which restrict the size of parts that can be printed. Anything bigger will need to be printed in separate parts and joined together after production. This can increase costs and time for larger parts due to the printer needing to print more parts before manual labor is used to join the parts together.

3. Post Processing

Although large parts require post-processing, as mentioned above, most 3D printed parts need some form of cleaning up to remove support material from the build and to smooth the surface to achieve the required finish. Post processing methods used include water jetting, sanding, a chemical soak and rinse, air or heat drying, assembly and others. The amount of post processing required depends on factors including the size of the part being produced, the intended application and the type of 3D printing technology used for production. So, while 3D printing allows for the fast production of parts, the speed of manufacture can be slowed by post processing.

4. Part Structure

In Additive Manufacturing parts are produced layer-by-layer. Although these layers adhere together it also means that they can delaminate under certain stresses or orientations. This problem is more significant when producing items using fused deposition modelling (FDM), while polyjet and multijet parts also tend to be more brittle. In certain cases it may be better to use injection moulding as it creates homogenous parts that will not separate and break.

5. Design Inaccuracies

Another potential problem with Additive manufacturing is directly related to the type of machine or process used, with some printers having lower tolerances, meaning that final parts may differ from the original design. This can be fixed in post processing, but it must be considered that this will further increase the time and cost of production.

1.6. Applications of Additive Manufacturing

Initially seen as a process for concept modeling and rapid prototyping, AM has expanded over the last five years or so to include applications in many areas of our lives. From prototyping and tooling to direct part manufacturing in industrial sectors such as architectural, medical, dental, aerospace, automotive, furniture and jewelry, new and innovative applications are constantly being developed. It can be said that AM belongs to the class of disruptive technologies, revolutionizing the way we think about design and manufacturing. From consumer goods produced in small batches to large scale manufacture, the applications of AM are vast. The number of users of these technologies has been growing constantly, from artists, designers and individuals to large companies and enterprises using AM to manufacture a wide range of final products [24].

1. *Medical*

Since the 1990s, Bio printing has been a staple in additive manufacturing that is directly applied to the medical field. Since the first break through in 1999 with a 3D printed human liver, there are many companies who have become the pioneers of this technology, such as Organovo, the creators of the first commercial bio printer, the NovoGen MMX, as well as different universities, and companies that have machine models to meet this new trend. Today, the technology has continued to grow and innovate into a strong and vibrant part of the 3D printing world, which is expected to grow by 36% by the year 2022. In addition to bio printing, another big sector of additive manufacturing in health is prosthetics. Thanks to the customizability that 3D printing offers, the impact that it has had on this industry has been phenomenal. The ability to easily transport 3D printers has made it a key tool in conflict zones to help those who have lost limbs a typical example is shown in the figure 1.11 given below, with MakerBot being one of the leaders in this area of healthcare. Additionally, a growing trend is being seen in hospitals around the world that are now incorporating 3D printers into their facilities to help meet the needs of their patients such as through additive manufacturing prosthetics as well as turning to 3D printing for the creation of 3D printed organs to practice on before operating on patients [25].

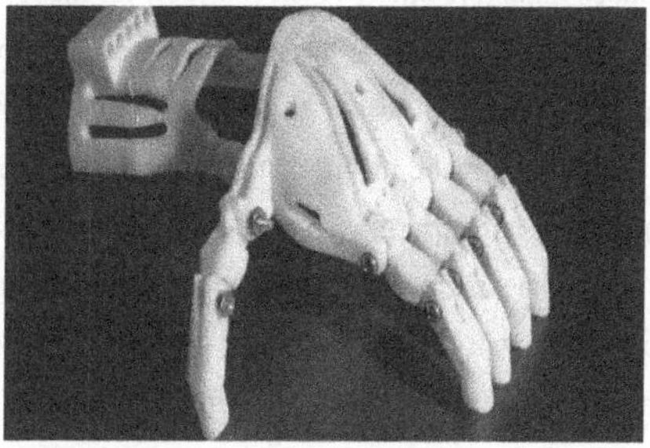

Fig. 1.11: 3D Printed Prosthetic Arm [26]

2. *Automotive*

AM has been deployed in automotive industry for decades, however it is usually used for the production of prototypes, or complex custom parts for luxurious or antique cars [27]. For specialty cars or low-volume production, AM can be economical for some parts. Applications include custom parts on luxury cars or replacement parts on antique cars. For example of Bentley Motors used AM to fabricate some custom interior components, such as bezels, that were subsequently covered in leather and other materials.

Typically, Bentley has production volumes of less than 10,000 cars for a given model, so this qualifies as low production volume. Companies and research institutes also have successfully applied AM techniques to manufacture functional components for racing vehicles. Vehicles for motorsports usually use lightweight alloys (e.g., titanium) and have highly complex structures and low production volumes [28], thus they may use AM more than passenger car industries. Figure 1.12 shows an example for commercially produced parts for automotive industries.

Fig 1.12: Example for Components Manufactured by AM Technique [29]

3. Aerospace

Due to the fact that aerospace companies are project oriented manufacturers, a lot of different components are produced in low quantities [30], thus the use of AM becomes more attractive in these aerospace industry. Since AM isn't suitable for mass production but are more convenient in small scale production. Just as in automotive applications the most important factor in aerospace applications is to consider the decrease of the automobile weight, while maintain its safety [31]. Current applications of AM in aerospace industry includes printing low volume complex components and spare parts in metals and plastics [32].Using AM to manufacture the insides of some parts can reduce their weight by up to 70 percent, which can save an airline millions of gallons of fuel every year. For example, General Electric jet engines used to have 20 separate pieces of metal welded together, to achieve a shape that efficiently mixes fuel and air in a fuel injector as shown in figure 1.13.

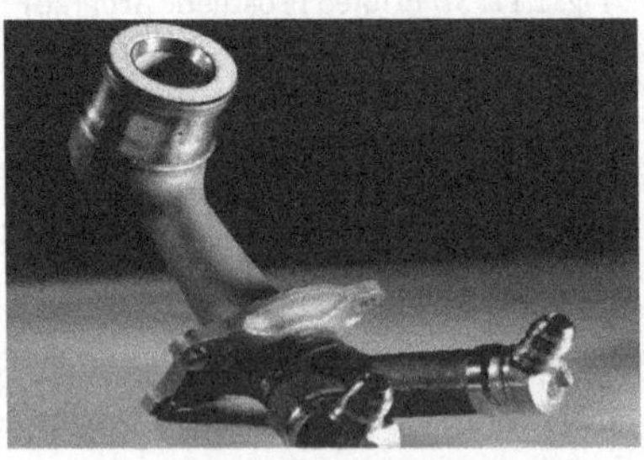

Fig. 1.13: Additive Fuel Nozzle [33]

4. *Tooling*

Tools, dies, and molds, usually referred to as "tooling," are required to produce nearly all plastic and metal products in industries such as automotive, medical, aerospace, and consumer electronics. Nevertheless, tooling production is a time consuming, technically difficult, and expensive production process that requires specialized materials, labor, and manufacturing techniques. Despite the expense and difficulty of designing and manufacturing tooling, dies and molds often have limited in-service lifetimes due to harsh use conditions including large loads under elevated temperatures, thermal cycling, repeated/ reversed loading conditions, contact with corrosive materials, and continuous production schedules. Thus AM techniques have been utilized in this industry for decades, most common techniques used in manufacturing is Selective Laser Sintering (SLS), Direct Metal Deposition (DMD), and Fused Deposition Modeling (FDM) [34]. A typical example for a die manufactured by AM process is shown in figure 1.14.

Fig. 1.14: Additive Manufacturing in Tooling [35]

5. *Construction*

The application of additive manufacturing (AM) in construction has been increasingly studied in recent years. Large robotic arm- and gantry-systems have been created to print building parts using aggregate-based materials, metals, or polymers. Significant benefits of AM are the automation of the production process, a high degree of design freedom, and the resulting potential for optimization [36]. A real time example of the use of AM is shown in figure 1.15.

Fig. 1.15: Expansion of the Lewis Grand Hotel in the Philippines [37]

CHAPTER 2

2. Applications of Additive Manufacturing

2.1. Classification of Additive Manufacturing

The Better way to classify AM system is broadly by the initial form of its material, all AM Systems can be easily categorized into [38].

1) Liquid Based
2) Solid Based
3) Powder Based

The American Society for Testing and Materials (ASTM) group "ASTM F42 - Additive Manufacturing" formulated a set of standards that classify the range of Additive Manufacturing processes into 7 categories as shown in the figure 2.1 [39].

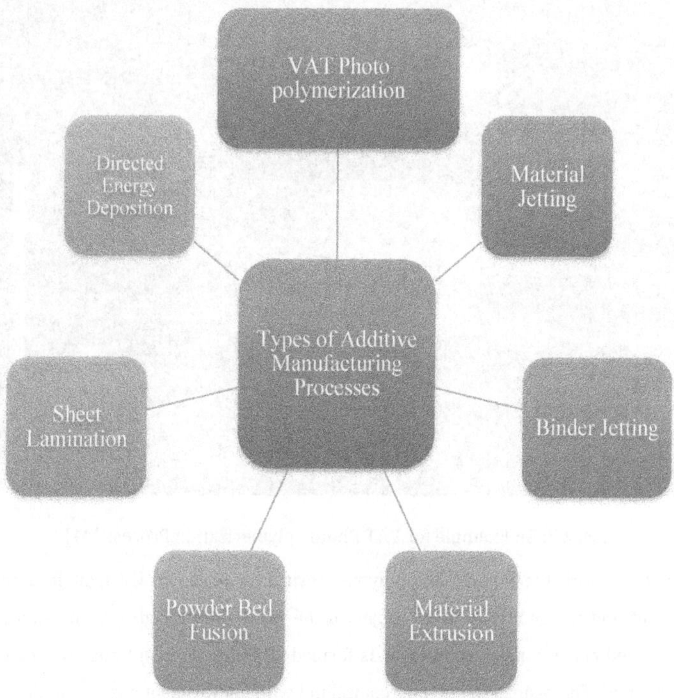

Fig. 2.1: Types of Additive Manufacturing Processes [38]

2.2. VAT Photo Polymerization

Vat photo polymerization, is a process by which a liquid is cured by a light source, turning it into a solid. This phenomenon known as polymerization and was the very first industrial additive manufacturing technology. SLA was developed in the early 1980s by Chuck Hull, founder of 3D Systems, who then patented the process and took it to market in 1987 with the SLA-1 machine.

2.2.1. Working Principle

Vat Photo polymerization uses a vat of liquid photosensitive polymer resin. This resin hardens on exposure to UV light. This property is used to build objects layer-by-layer [40]. An example for VAT Photo polymerization process is shown in the figure 2.2.

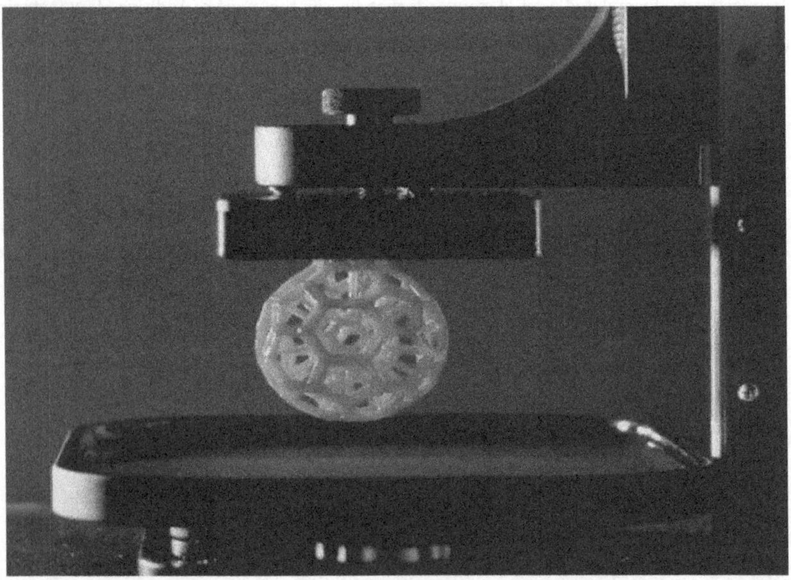

Fig. 2.2: An Example for VAT Photo Polymerization Process [41]

A vat filled with the liquid photopolymer resin is exposed to UV light in a controlled environment and the geometry of the object to be printed is traced out. The exposed resin hardens (called curing) and a solid layer is formed. This process continues till the complete object is printed. The resins are polymer compound with additives for specific applications like Tough, Flexible, Dental, etc. These processes impart high quality surface finish to the object. Most common processes in this category are shown in the figure 2.3.

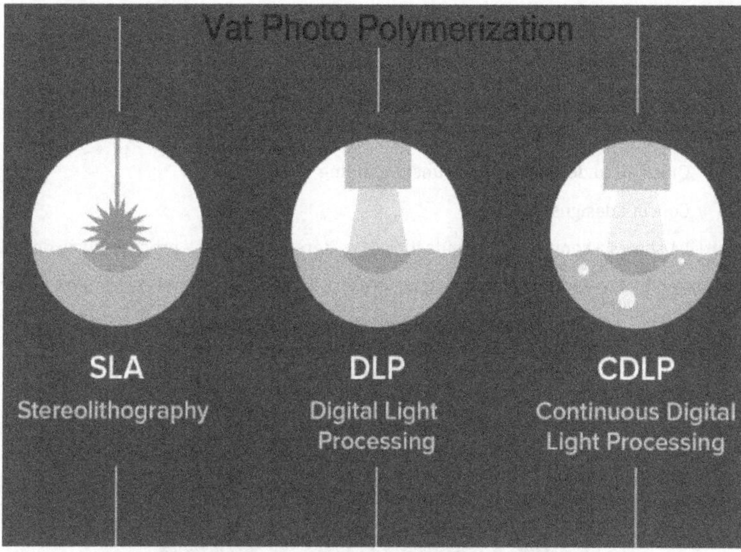

Fig. 2.3: Types of Processes in VAT Photo Polymerization [42]

2.2.2. Materials Used

The photopolymer resins used in vat photo polymerization processes are usually either epoxy- or acrylate-based. While epoxy-based resins require more energy to solidify, they are currently the most used in industrial applications due to their better mechanical and thermal properties. Epoxy resins are also less prone to shrink during processing than acrylate-based resins. On the other hand, acrylate-based resins are very versatile, ranging from extremely tough to flexible, rubber-like materials. These resins are also more color-stable, so they're more often used when a specific color is needed [43].

2.2.3. Applications

Stereolithography the most commonly used VAT photo polymerization process is a very versatile method with applications in a variety of industries. The aerospace and automotive industries can, for instance, benefit from rapid manufacturing of high-performance materials. Microfluidics and medicine are furthermore, significant fields where SLA shows great potential and is already being applied successfully [44]. Applications in medicine, where patient-specific designs are often necessary to accommodate for individual anatomies, can greatly benefit from AM as well. Apart from the medical sector and automobile industry, use of this technique is also in the manufacture of the following objects such as [45]:

- Optical lenses
- Design evaluation
- Scale designs
- Masters for vacuum molding
- Quick Mold designs for investment molding
- Concept designs

Figure 2.4 shows a component manufactured by Stereolithography process which describes the role played by the additive manufacturing process in medical field in replacement of parts.

Fig. 2.4: Component Manufactured by Stereolithography Process [44]

2.3. Material Extrusion

Material extrusion is an additive manufacturing technique which uses continuous filament of thermoplastic or composite material to construct 3D parts. The material in the form of plastic filament fed through an extruding nozzle, where it heated and then deposited onto the build platform layer by layer.

2.3.1. Working Principle

Material Extrusion 3D printing technology uses a continuous filament of a thermoplastic material as a base material. The filament is fed from a coil, through a moving heated printer extruder head, often abbreviated as an extruder. The molten material is forced out of the extruder's nozzle and is deposited first onto a 3D printing platform, which can be heated for extra adhesion [46]. Once the first layer is completed, the extruder and the platform are parted away in one step, and the second layer can then be directly deposited onto the growing workpiece. The extruder head is moved under computer control. At least three axes are required for the extruder to move in Cartesian architectures, but polar and delta systems are also becoming increasingly popular. One layer is deposited on top of a previous layer until the object's fabrication is complete. Figure 2.5 shows a graphical illustration of a material extrusion process.

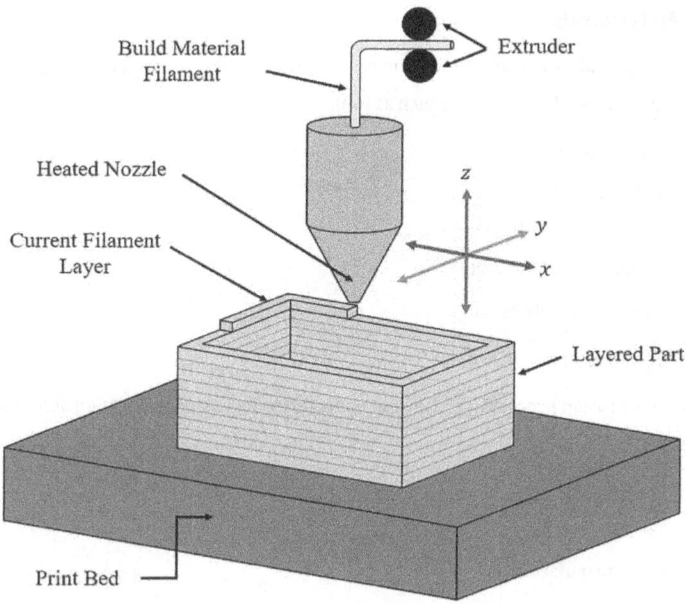

Fig. 2.5: Schematic Representation of Material Extrusion Process in AM [47]

Whenever necessary, support structures are included in the process to enable the fabrication of complex geometrical features. This basic principle enables the production of complex parts without a shaping tool other than a die with a simple geometry, generally round. Depending on the type of extruder used, one can classify material extrusion additive manufacturing into different types as shown in the figure 2.6 given below.

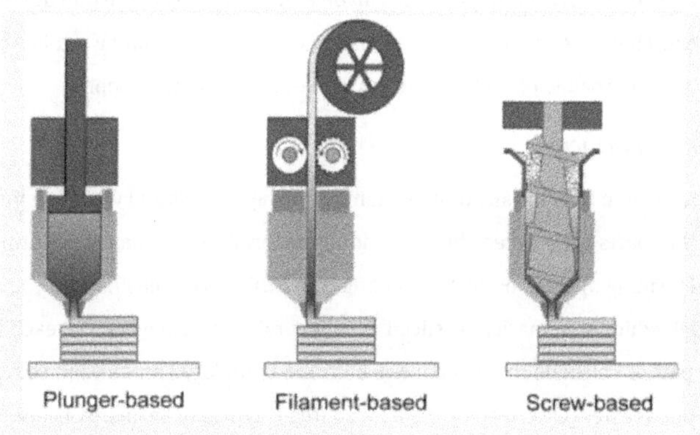

Fig. 2.6: Different Types of Material Extrusion based AM Process [48]

2.3.2. *Materials Used*

Several materials are available for this process. ABS is the most widely used polymer, but other polymers have also been used such as [49]:

- Polycarbonate
- High density polyethylene
- Polycarbonate/ABS blends
- Polyphenylsulfone, and
- High impact polystyrene

2.3.3. *Applications*

Material extrusion process in additive manufacturing has a vast application across different field as listed below [50].

- Prototyping
- Tooling
- Office manufacturing
- Automotive
- Aerospace
- Medical field
- Functional testing

2.4. Powder Bed Fusion

Powder bed fusion (PBF) methods use either a laser or electron beam to melt and fuse material powder together. All PBF processes involve the spreading of the powder material over previous layers. There are different mechanisms to enable this, including a roller or a blade. A hopper or a reservoir below of aside the bed provides fresh material supply.

2.4.1. *Working Principle*

Powder bed fusion (PBF) is an additive manufacturing process and works on the same basic principle in that parts are formed through adding material rather than subtracting it through conventional forming operations such as milling. The PBF process begins with the creation of a 3D CAD model, which is numerically 'sliced' into several discrete layers. For each layer, a heat source scan path is calculated which defines both the boundary contour and some form of fill sequence, often a raster pattern since the heat source is typically an energy beam. Each layer is then sequentially bonded on top of each other.

PBF processes spread powdered material over the previously joined layer, ready for processing of the next layer hence the manufacturing is discrete rather than continuous (though each layer is fully consolidated to adjacent layers). A hopper supplies the powdered material which is then spread uniformly over the powder bed build platform area via a roller or blade. The optimal thickness of each layer of spread powder is dependent on the processing conditions and material used, but values of 25 to 100μm are common. A graphical illustration of the working methodology of the process is shown in the figure 2.7.

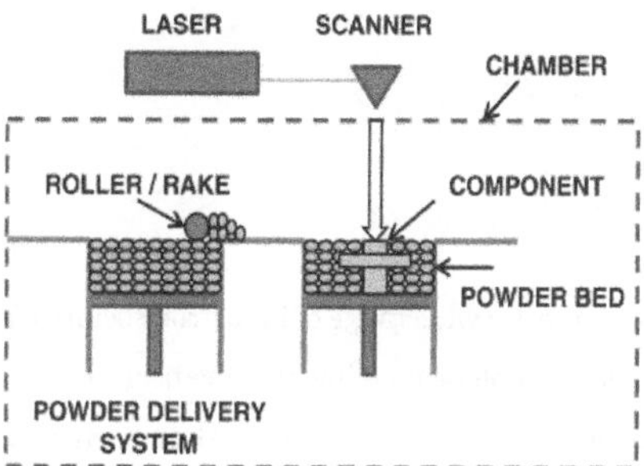

Fig. 2.7: Powder Bed Fusion Process [51]

There are several variants of PBF, which are designated by the heat source used and the type of material joined. The different variants are indicated in figure 2.8.

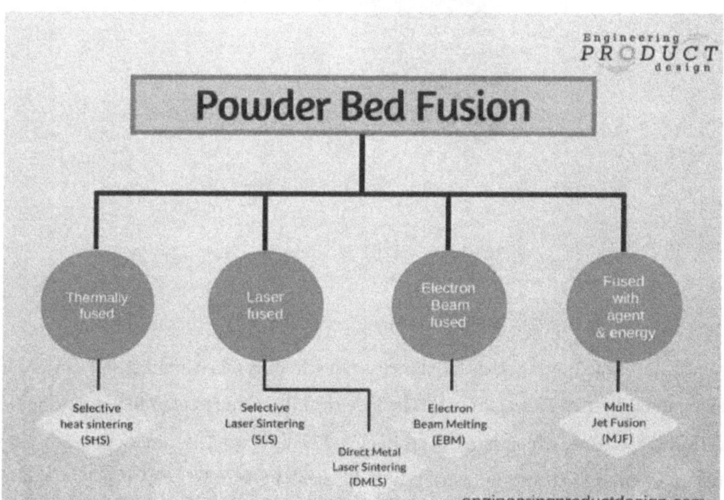

Fig. 2.8: Different Variants in Powder Bed Fusion Process [52]

2.4.2. *Materials Used*

The Powder bed fusion process uses any powder based materials, but common metals and polymers used are as listed below [53].

- Polyamide (Nylons)
- Alumide (a blend of gray aluminum powder and polyamide)
- Aluminum
- Titanium-based alloys
- Stainless Steel
- Cobalt Chrome
- Copper

2.4.3. *Applications*

PBF processes are used across a wide range of industrial sectors for numerous applications. For example, the process is implemented by the medical sector for making customized orthopedic components, such as titanium alloy cranial or acetabular implants as shown in figure 2.9.

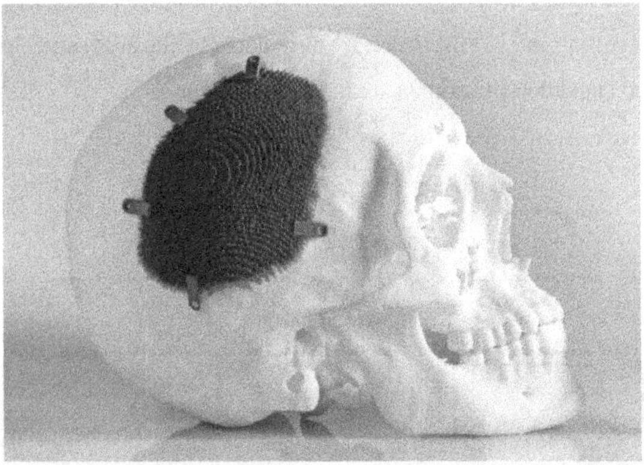

Fig. 2.9: Cranial Implant Manufactured by PBF Process [54]

From an aerospace perspective, PBF processes are finding much interest and use for military and commercial aircraft. Examples of this include the PBF manufactured fuel nozzle on General Electric's GE9X engine, which is used on Boeing 777 aircraft. The production use cases are mostly found in Aerospace, Industrial, and Medical industries, where low volumes and high-value geometries can justify its high cost and complexity.

2.5. Material Jetting

Material jetting creates objects in a similar method to a two dimensional ink jet printer. Material is jetted onto a build platform using either a continuous or Drop on Demand (DOD) approach. The material layers are then cured or hardened using ultraviolet (UV) light. Material Jetting stands out among other 3D printing technologies for its ability to produce highly accurate parts with a smooth surface finish. Since its emergence in the late 1990s, Material Jetting is an ideal 3D printing technology for producing full-colour, visual prototypes, injection moulds and casting patterns [55].

2.5.1. Working Principle

Material Jetting is an inkjet printing process whereby printheads are used to deposit a liquid photo reactive material onto a build platform layer upon layer. Similarly to Stereolithography (SLA), Material Jetting uses a UV light to solidify the material. The methods of material deposition vary from printer to printer and can involve either a continuous or Drop-on-Demand (DOD) jetting approach. Figure 2.10 shows a schematic representation of material jetting process. Once the liquid resin is preheated for better viscosity, the print head begins to move above the build platform, depositing the first layer of material where required. Support structures are printed simultaneously with the part to ensure its stability during printing. The deposited material is then exposed to UV light, curing (solidifying) the layer of material in a process known as photo polymerization. Once the first layer has solidified, the build platform is lowered by one layer height and the process is repeated until the part is finished.

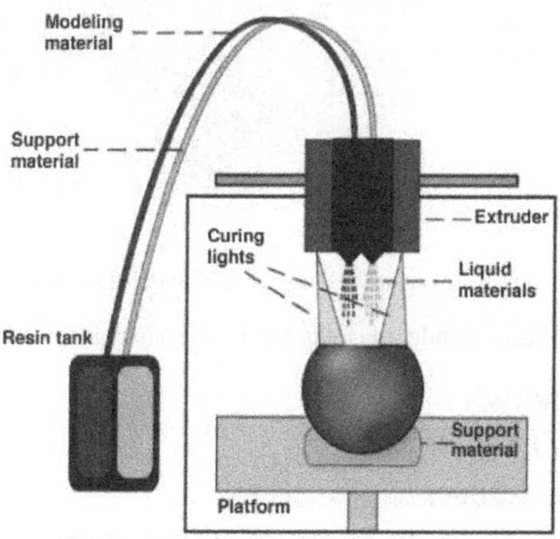

Fig. 2.10: Schematic Representation of Material Jetting Process [56]

One of the biggest advantages of this technology is the ability to gang multiple print heads together. Having multiple print heads allows these machines to do unique things, such as print in different colors like traditional ink-jet printers, print faster by printing over the entire build surface in one pass, and print in multiple materials at the same time. The surface quality of these parts is usually quite high due to jetting very small droplets. Similar to how a normal inkjet printer is able to print thousands of different colors using only three different inks, a 3D printer that is able to jet multiple materials can combine these materials in different proportions in order to vary material properties in the finished part and create so-called digital materials.

2.5.2. Types of Material Jetting

Since the inception of Material jetting technology in the 1990s, several techniques have emerged and progressed into a developed state. Although they differ from manufacturer to manufacturer on how the material is cured and how they are deposited, the underpinning material jetting technique is the same.

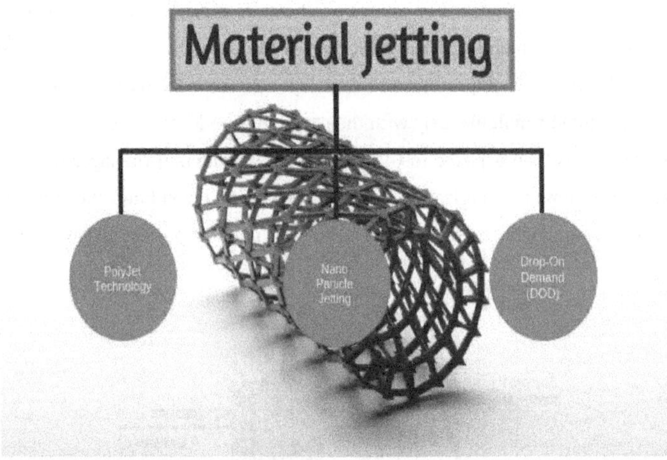

Fig. 2.11: Types of Material Jetting Process [57]

They all can be categorized under the following three main types as illustrated in figure 2.11.

- Poly Jet technology
- Nano Particle Jetting (NPJ)
- Drop-On Demand (DOD)

Poly Jet technology, ultra-thin layers of photopolymer material onto the build tray, very similar to a standard inkjet printer and each layer is immediately cured by a UV light.

The gel-like support material is available for complex geometrical parts which can be easily removed by hand or by using an aqua blast or water jetting. A schematic representation of poly jet printing process is shown in figure 2.12.

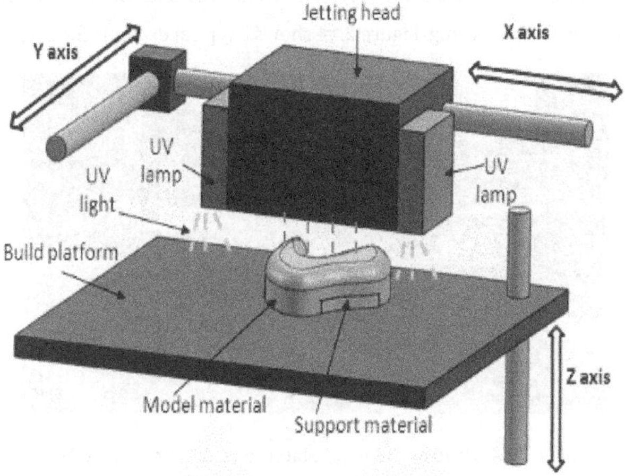

Fig. 2.12: Schematic Representation of Poly Jet Printing Process [58]

Nano Particle Jetting technology uses a unique dispersion methodology to deposit the material onto the build platform. Special solid nanoparticles of the build material are contained or suspended in a liquid and then are jetted onto the build tray. Both build and support materials which are suspended in liquid form are supplied in sealed cartridges. Once the build process is complete, the support material is removed before going through a simple sintering process. The figure 2.13 shows some examples of ceramic components manufactured by Nano Particle Jetting process.

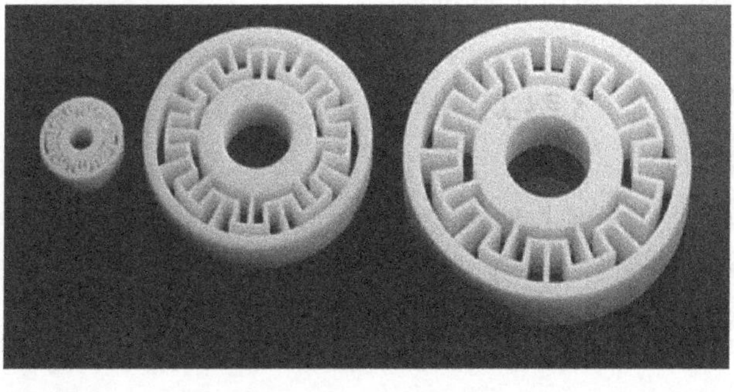

Fig. 2.13: Ceramic Components Manufactured by Nano Particle Jetting [59]

Drop on Demand as the name implies deposits only tiny dots of material when necessary instead of continuous lines and used typically for viscous liquid materials. These DOD technology printers use two heads one for each build and support material. Due to its ability to print viscous liquid, these are used exclusively to make wax patterns for investment casting, especially jewellery manufacturing. Figure 2.14 shows a typical example for DoD processes.

Fig. 2.14: Drop on Demand Material Jetting Process [60]

2.5.3. *Material Used*

At the beginning, MJ materials were waxes and some photopolymers, but nowadays, the variety is huge thanks to mixing (blending) of the aforementioned materials in differing combinations [61]. The material jetting process uses polymers and plastics. Some of the commonly used materials are listed below.

- Polypropylene.
- HDPE - High Density Poly Ethylene.
- PS - Polystyrene.
- PMMA – Poly (methyl methacrylate).
- PC - Polycarbonate
- ABS - Acrylonitrile butadiene styrene

2.5.4. *Applications*

- Prototypes are built by material jetting are used for visual and form / fit testing.
- Casting patterns are produced with very good accuracy and surface finishes.
- Used in the medical, dental and jewelry industry.
- Fabrication of multi-material parts.
- Manufacturing realistic components.

- Tooling.
- Injection mold manufacturing.
- Multi-material 3D printing technology is often used in the production of 3D printed prosthetics.

2.6. Binder Jetting

In Binder Jetting, a binder is selectively deposited onto the powder bed, bonding these areas together to form a solid part one layer at a time. The materials commonly used in Binder Jetting are metals, sand, and ceramics that come in a granular form.

2.6.1. Working Principle

A Binder Jetting 3D printer is normally composed of two tanks and a printing platform. At the beginning of the printing process, one of the tanks is empty, whilst the other holds the powdered printing material. The leveling roller spreads a first layer of powder over the platform. The print head, similar to those of 2D printers sprays the binding agent in the form of droplets of 80μm. As a result, the powder is unified and the piece is formed, layer after layer. It is possible to print in color using this technology. For example, in the case of ceramics, the color is added at the same time as the binder. As the binder and color is added to the successive layers of powder, the piece is formed in full-color [62]. Figure 2.15 illustrates a schematic representation of binder jetting process.

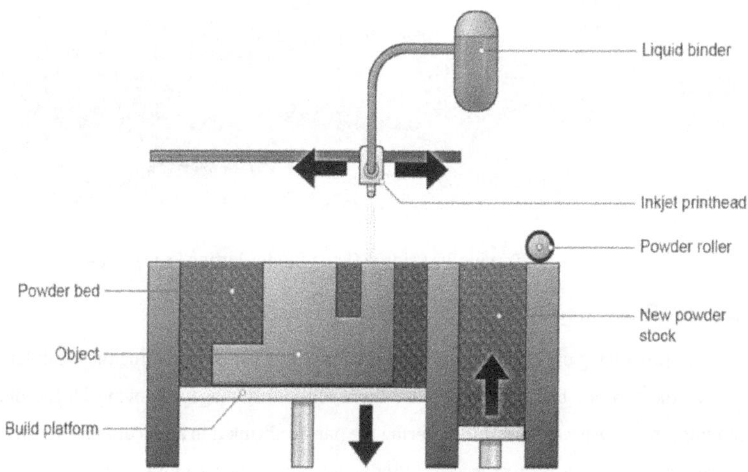

Fig. 2.15: Schematic Representation of Binder Jetting Process [63]

Once printing is complete, the pieces must be freed from the unsolidified powder that covers them. They also need to be cured to increase their resistance and revive the colors. The powder residue is removed with a brush or using pressurized air. Depending on the material used, be it ceramics or metals, a further step is necessary. In the case of metal Binder Jetting, the pieces must be sintered or infiltrated with a low-melting-temperature metal such as bronze. In the case of colored ceramic pieces, it is also necessary to cover them by adding acrylic in order to improve the colors and mechanical resistance of the pieces. Figure 2.16 shows the practical application of binder jetting process used for manufacturing M5 screws.

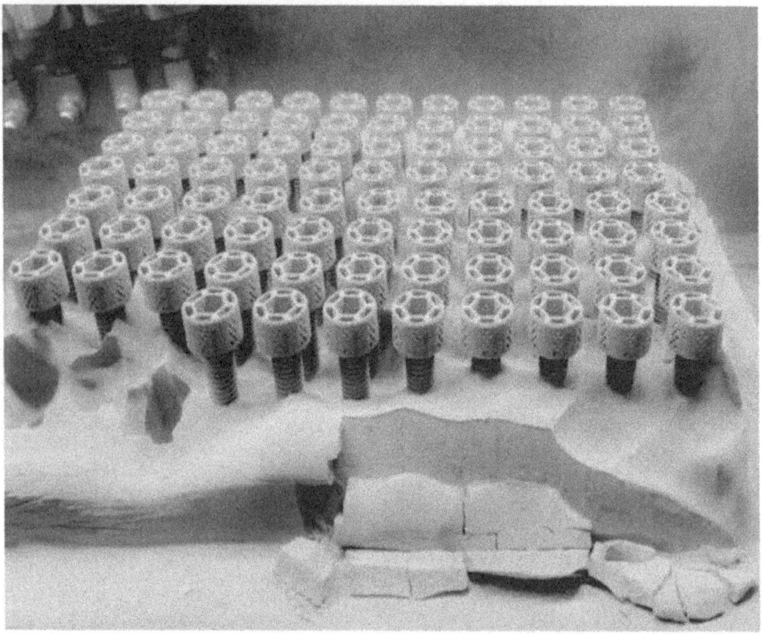

Fig. 2.16: The M5 Screws 3D Printed Using Binder Jetting Process [64]

2.6.2. *Post Processing Steps in Binder Jetting Process*

The post-processing of binder-jet printed parts has three goals: extracting the weak "green" parts from the powder bed and de-powder them without damage, removing all the binder (de-binding) from the green part, and sintering the part to shrink it to an acceptable density and geometric accuracy. The post-processing Binder Jet parts can be more important than the printing process itself. The step involved in post processing is illustrated in figure 2.17.

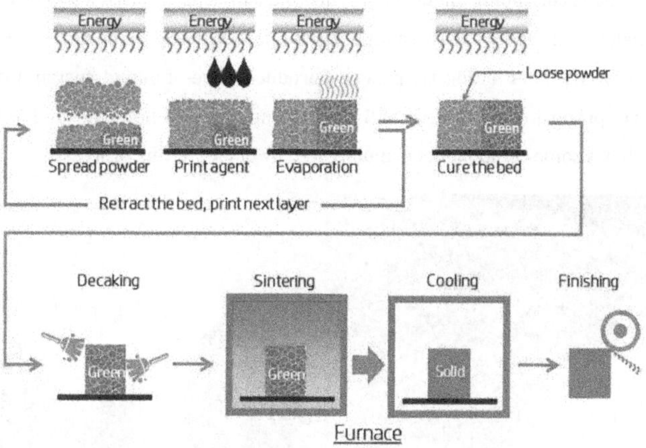

Fig. 2.17: Post Processing Steps in Binder Jetting Processes [65]

- **Curing and sintering:** This involves the application of heat, usually in a controlled environment like a furnace, to effectively fuse the particles together.
- **Sintering:** The binder agent is burned away in this process, leaving the part quite porous.
- **Infiltration:** To reduce porosity and increase density, the parts are infiltrated with different materials that fill these voids.
- **Finishing:** By the end of post-processing, the parts display much higher resistance and better mechanical properties.

2.6.3. Material Used

Binder jetting is widely applied for ceramic, metallic, and even sand parts. Regardless of the material, all objects require post-processing.

2.6.4. Applications

Binder Jetting has a wide range of applications, including:

- Casting patterns
- Cores and molds
- Aerospace parts
- Prototypes
- Full color decorative objects
- Jewellery

Binder Jetting is commonly applicable in such industries as aerospace and heavy industry, where thermally stable and wear-resistant parts are needed. Jewelry and decorative industries have also embraced Binder Jetting as a means of producing one-of-a-kind, custom-made objects. An unusual yet promising application of Binder Jetting may also lie in the food industry [66]. Figure 2.18 shows some components manufactured by binder jetting process.

Fig. 2.18: Example of Components Manufactured by Binder Jetting Process [67]

2.7. Direct Energy Deposition

Directed Energy Deposition (DED) is a 3D printing method which uses a focused energy source, such as a plasma arc, laser or electron beam to melt a material which is simultaneously deposited by a nozzle. As with other additive manufacturing processes, DED systems can be used to add material to existing components, for repairs, or occasionally to build new parts. The DED process is known by other names, including Laser Engineered Net Shaping (LENS), Direct Metal Deposition (DMD), Electron Beam Additive Manufacturing (EBAM), Directed Light Fabrication, and 3D Laser Cladding, depending on the exact application or method used.

2.7.1. Working Principle

Directed Energy Deposition works by depositing material that has been melted onto a specified surface where it solidifies, fusing materials together to form a structure. DED machines typically use a nozzle that is mounted on a multi axis arm which can move in multiple directions, allowing for variable deposition. The process is typically performed within a controlled chamber with reduced oxygen levels. With electron beam-based systems the process is performed in a vacuum, while laser-based systems use a fully inert chamber when working with reactive metals. It is also possible to use a shielding gas to shroud the part and prevent contamination during metal 3D printing [68].

DED uses a heat source to melt a powder or wire as it is deposited onto the surface of an object. While powder provides greater accuracy in deposition, wire is more efficient with regards to material use. The material is added layer-by-layer and solidifies from the melt pool to create new features. A schematic representation of direct energy deposition method is as shown in figure 2.19. Layers are typically 0.25mm to 0.5mm thick. The cooling times for materials are very fast at around 1000-5000 °C per second. The cooling time affects the final grain structure although overlapping in the material can cause re-melting, which creates a uniform but alternating microstructure.

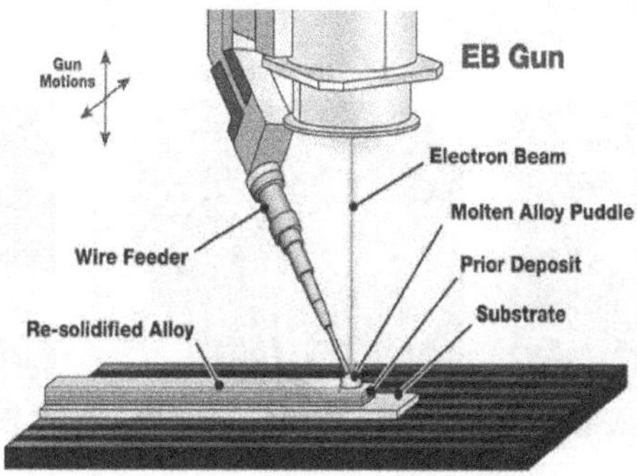

Fig. 2.19: Schematic Representation of Direct Energy Deposition Process [69]

In most cases, the object remains in a fixed position while the arm moves to lay down the material. However, this can be reversed with the use of a platform, which moves while the arm remains stationary.

2.7.2. Materials Used

Predominantly, DED uses metal in the form of either wire or powder. However, DED technology is also capable of using polymers and ceramics. DED supports a wide range of metals as any metal that can be welded can be used to create parts through this technology. Almost any weldable metal can be additively manufactured using DED. Commonly used materials by various manufacturers are as follows [70];

- Aluminium
- Inconel
- Niobium
- Stainless steel
- Tantalum
- Titanium

- Titanium alloys
- Tungsten

2.7.3. *Applications*

- Fixing mistakes
- Repair and restoration
- Multi material applications

A practical industrial application of gear teeth repair using Direct Energy Deposition method is shown below in figure 2.20.

Fig. 2.20: Illustrates How DED Method is Used for Repair of Gear Teeth [71]

2.8. Sheet Lamination

Sheet lamination is an additive manufacturing (AM) methodology where thin sheets of material (usually supplied via a system of feed rollers) are bonded together layer-by-layer to form a single piece that is cut into a 3D object. Laminated object manufacturing (LOM) and ultrasonic consolidation (UC) are both examples of sheet lamination techniques. Sheet Lamination is a low temperature process and is used for making models for display. The process is quite useful for making visual models speedily at low cost.

2.8.1. *Working Principle*

Sheet lamination processes include ultrasonic additive manufacturing (UAM) and laminated object manufacturing (LOM). The Ultrasonic Additive Manufacturing process uses sheets or ribbons of metal, which are bound together using ultrasonic welding. The process does require additional cnc machining and removal of the unbound metal, often during the welding process. Laminated object manufacturing (LOM) uses a similar layer by layer approach but uses paper as material and adhesive instead of welding. A simple schematic representation of sheet lamination process is shown in figure 2.21.

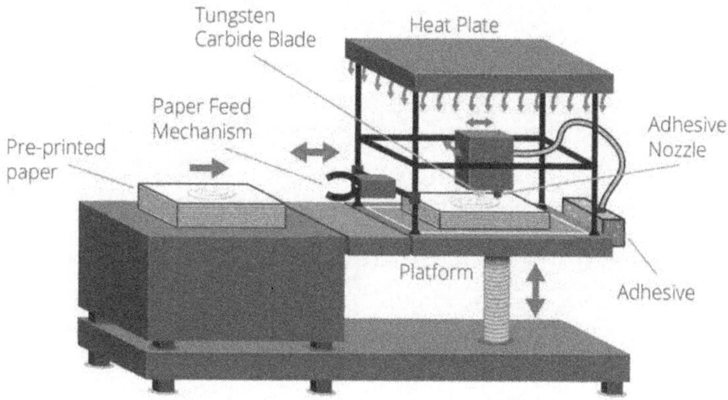

Fig 2.21: Schematic Representation of Sheet Lamination Process [72]

The various steps involved in the process are as follows.

- The material is positioned in place on the cutting bed.
- The material is bonded in place, over the previous layer, using the adhesive.
- The required shape is then cut from the layer, by laser or knife.
- The next layer is added
- Steps two and three can be reversed and alternatively, the material can be cut before being positioned and bonded.

2.8.2. *Materials Used*

Although sheet lamination can use a wide variety of materials, such as paper, polymer, ceramic and metal, each material uses different binding methods. Most common sheet lamination material is paper with pre applied adhesive where heat and pressure are used to activate the adhesive lay [73]. Polymers use heat and pressure without the adhesive as it relies on melting the sheets together. Metal sheets are bound using ultrasonic welding, while materials such as fibre-based material and ceramic use thermal energy in the form of oven baking to combine the layers together.

2.8.3. *Applications*

- Joining dissimilar metals
- Rapid prototyping
- Colorful prototyping

CHAPTER 3

3. Fused Deposition Modelling

3.1. Introduction to Fused Deposition Modeling [FDM]

The FDM process, which is classified as the material extrusion AM technique, is the simplest, affordable and readily available 3D printing technique for polymer-based materials and it has been extensively used in various industries a simple illustrative diagram of the FDM process is shown in figure 3.1.

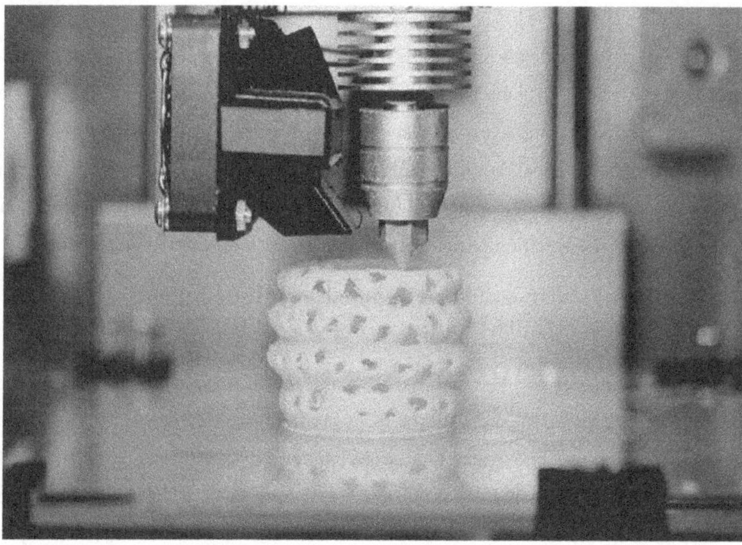

Fig. 3.1: Illustrative Diagram of FDM [74]

Fused Deposition Modeling (FDM) is a 3D printing technique pioneered in the 1990s by Stratasys. There are several different methods of 3D printing but, FDM has become the most widely used 3D printing method in the world. Fused deposition modelling (FDM), also known as the material extrusion additive manufacturing technique, utilizes polymers as the raw material (filament). The filament is usually heated to a molten state and then extruded through the nozzle of the machine (3D printer). Figure 3.2 shows the market size of each 3D printing process and the expected growth by 2024.

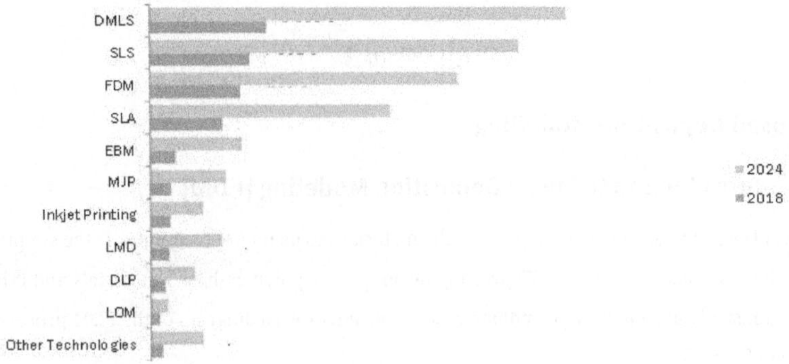

Fig. 3.2: Illustrates the Market Size of Various 3D Printing Processes [75]

In Fused deposition modeling process the filament is continuously fed through the extruder and nozzle of the machine via the two rollers rotating in opposite directions. The material is deposited on the build plate layer-by-layer until the required product shape and size are achieved. During the layering, the printer nozzle navigates back and forth as per the spatial coordinates of the original CAD model until the desired size and shape of the component is produced. In some FDM systems multiple extrusion nozzles can be used for deposition.

3.2. Working Principle of Fused Deposition Modeling

FDM begins with a software process, developed by Stratasys, which processes an STL file (Stereolithography file format) in minutes, mathematically slicing and orienting the model for the build process. If required, support structures are automatically generated. The machine dispenses two materials - one for the model and one for a disposable support structure. The thermoplastics are liquefied and deposited by an extrusion head, which follows a tool-path defined by the CAD file. The materials are deposited in layers as fine as 0.04 mm (0.0016") thick, and the part is built from the bottom up - one layer at a time. FDM works on an "additive" principle by laying down material in layers. A plastic filament or metal wire is unwound from a coil and supplies material to an extrusion nozzle which can turn the flow on and off. The nozzle is heated to melt the material and can be moved in both horizontal and vertical directions by a numerically controlled mechanism, directly controlled by a computer-aided manufacturing (CAM) software package. The model or part is produced by extruding small beads of thermoplastic material to form layers as the material hardens immediately after extrusion from the nozzle. Stepper motors or servo motor are typically employed to move the extrusion head [77]. Figure 3.3 shows a simple schematic representation of FDM process.

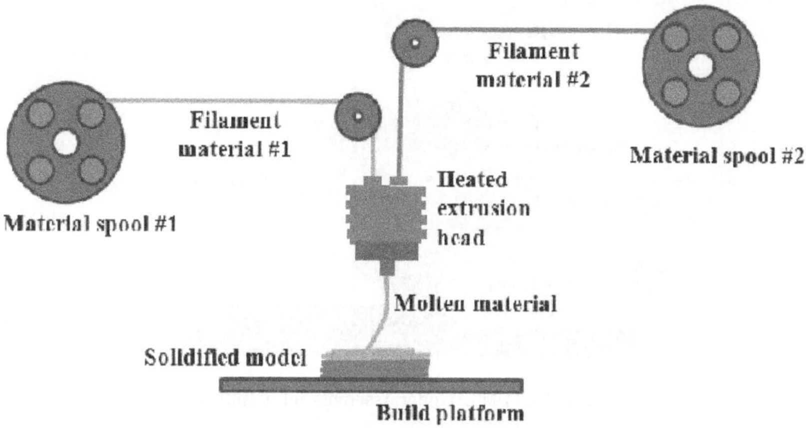

Fig. 3.3: Schematic Representation of FDM Process [78]

Several materials are available with different trade-offs between various properties such as strength and temperature. The most commonly used materials include Acrylonitrile butadiene styrene (ABS), Polycarbonates, Polycaprolactone, Polyphenylsulfone and waxes. A "water-soluble" material can be used for making temporary supports while manufacturing is in progress. Once an object comes off the FDM printer, its support materials are removed either by soaking the object in a water and detergent solution or, in the case of thermoplastic supports, snapping the support material off by hand. Objects may also be sanded, milled, painted or plated to improve their function and appearance.

3.3. Parts of a FDM 3D Printer

These components of FDM 3D printers help accomplish tasks from printing smaller to larger parts with ease. The list here contains the basic component of a FDM 3Dprinter [79].

- Extruder
- Structure
- Printing Surface
- Electronics

3.3.1. Extruder

The extruder is the "brush" of the 3D printer, one of the most important parts. You can differentiate two parts a nozzle that is heated call hot end, responsible for heating the material. And another part composed of a system of gears and bearing that pushes the filament and makes it pass through the nozzle. A typical FDM extruder assembly is shown in figure 3.4.

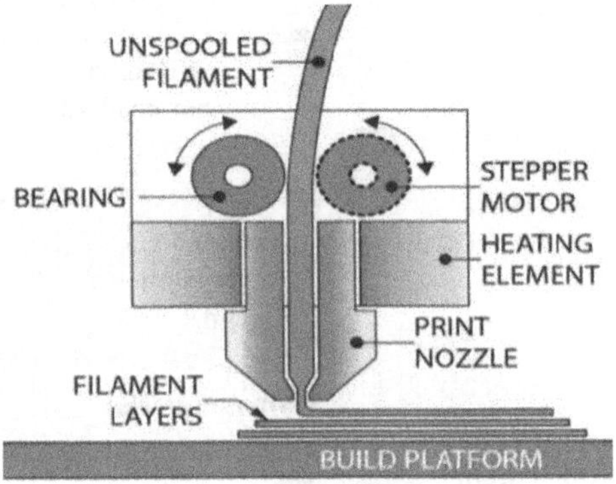

Fig. 3.4: FDM Extruder Assembly Setup [80]

The various components present in extruder are discussed below.

1. *Extruder Hotend*

The extruder hotend is one of the most important parts of a 3D printer. Its responsible for melting the filament and depositing it onto the print bed to form a solid object. The hotend counts for one-half of the print head or extruder, with the cold end being the other half. Ina direct drive setup, the hotend is positioned below the cold end, in a Bowden setup, the hotend is separate from the cold end which is mounted somewhere on the frame. Size-wise the hotend is one of the smallest parts of a 3D printer, but it has a big impact on the print quality and material compatibility [81]. A typical extruder hot end with a fan is shown in figure 3.5.

Fig. 3.5: Extruder hotend [82]

2. *Heat Sink*

A heat sink is an instrument that is used to lower the temperature of some components. In the extruder, the heat sinks are usually accompanied by a fan. Heat dissipation is an important problem in FDM. If the heat rises, it makes the filament softer on the top, which makes the task of feeding the filament to be a challenging one. An extruder hotend with an angled fan setup which is used as heat sink is shown in figure 3.6.

Fig. 3.6: Angled Fan Mount on Extruder Hotend [83]

3. *Thermal Bar/ Barrel*

It is thermal bar that plays a great role in guiding the filament to the tip. It makes the union between the Heat Block and the filament thrust system. Some use a Teflon tube (PTFE) to improve the sliding of the filament. Figure 3.7 shows a simple representation of a thermal bar.

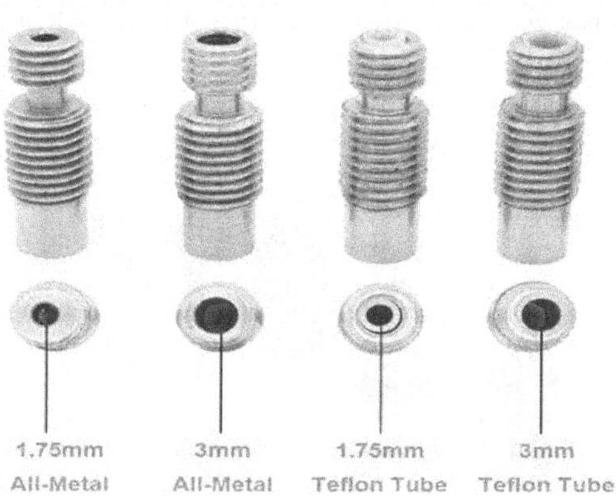

Fig. 3.7: Thermal Bar Used in FDM [80]

4. *Heating Block*

The heating block is the part responsible for heating and maintaining the heat. It makes of union between the mouthpiece and the thermal bar. To heat this block an electronic resistance is used. To know what temperature the block is in, a temperature sensor is used. Therefore the block has holes to insert both the resistance and the temperature sensor. The most common is that it is aluminium although you can find other materials. Each material has a coefficient of thermal dissipation. That is, each material has the property of maintaining heat. Figure 3.8 shows a heating block made of aluminium used in a fused deposition modeling process.

Fig. 3.8: Heating Block Made of Aluminium [84]

5. *Nozzle*

The nozzle gives the diameter with which the material will be extruded. As with the heater block, nozzles of different materials can be found. The most common is to use brass nozzles for the price. There are nozzles with different threads that adapt to different extruder models. With finer nozzles, you can make smaller details. With larger nozzles, you can print extrude more material in a single pass and print faster. Different types of brass nozzle used in FDM process is shown in figure 3.9.

Fig. 3.9: Brass Nozzles Used in FDM Processes [85]

3.3.2. *Types of Extruder*

Within extruders there are three types depending on the type of drive [86].

- Gear type extruder
- Direct Extrusion
- Bowden Extruder

1. *Gear Type Extruder*

The first extruders that were used were formed by a system of printed gears. In this way, it was possible to give more strength to the movement. It was mainly used for filaments with larger diameter. Figure 3.10 shows a schematic representation of gear type extruder used in FDM processes.

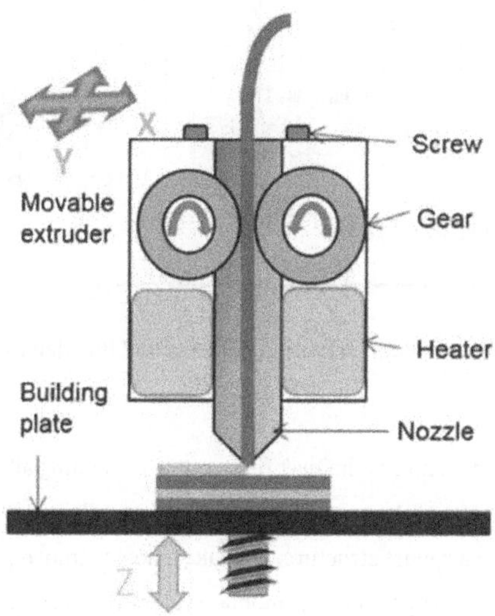

Fig 3.10: Schematic Representation of Gear Type Extruder [87]

2. *Direct Extrusion*

An evolution of the extruders with gears is those of direct extrusion. Direct extrusion refers to the motor pushing the filament directly. A toothed pulley or "Hobbed Pulley" is placed on the motor shaft. This pulley pushes the filament with the teeth. To ensure that the filament does not slip, a bearing is used on the opposite side to apply pressure. The main advantage of this type of extruder is that it occupies less space. Schematic representation of direct extruder is shown in figure 3.11.

3. *Bowden Extruder*

A Bowden extruder can carry both a direct extrusion system and gears. The difference is that a PTFE tube separates the drive system (motor) of Hot End. It is possible to reduce the weight considerably, making it possible to print at higher speeds. It also makes it easier to change the Hot End. An schematic illustration of Bowden extruder in shown in figure 3.11.

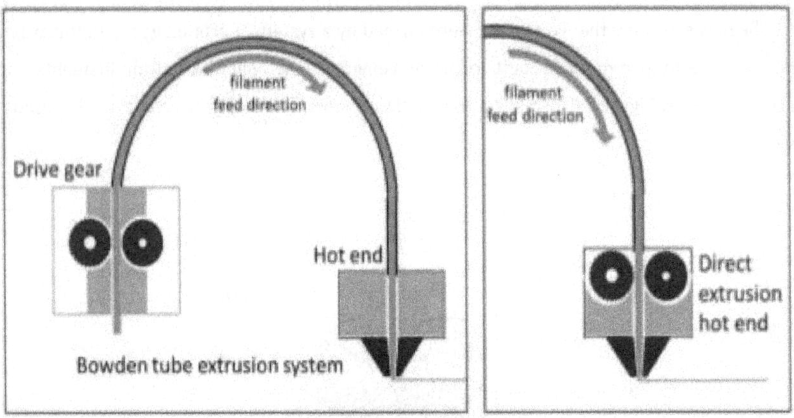

Fig. 3.11: Schematic Representation of Direct and Bowden Extrusion [88]

3.3.3. *Structure*

In FDM printing, the structure is the part responsible for joining all the parts of the printer and allowing the movements of the axes. Different structures allow different movements, but it is very important to have a robust structure. A weak structure transmits vibrations and forces printing at slow speeds. A 3D printer is a mobile system that meets certain characteristics. The extruder is always oriented downwards and can be moved with respect to the printing surface in the three axes (3 dimensions). The different types of structure are as discussed below [89].

- Cartesian
- Delta
- Polar
- Robotic arm

1. *Cartesian FDM 3D Printer*

Cartesian 3D printers are the most common FDM 3D printer found on the market. Based on the Cartesian coordinate system in mathematics, this technology uses three-axis: X, Y, and Z to determine the correct positions and direction of the print head. With this type of printer, the printing bed usually moves only on the Z-axis, with the print head, working two-dimensionally on the X-Y plane. Two well-known brands in the Fused Deposition Modeling market that use Cartesian technology for their FDM 3D printers are Ultimaker and MakerBot. Figure 3.12 shows a schematic representation of Cartesian type 3D printer used in industries.

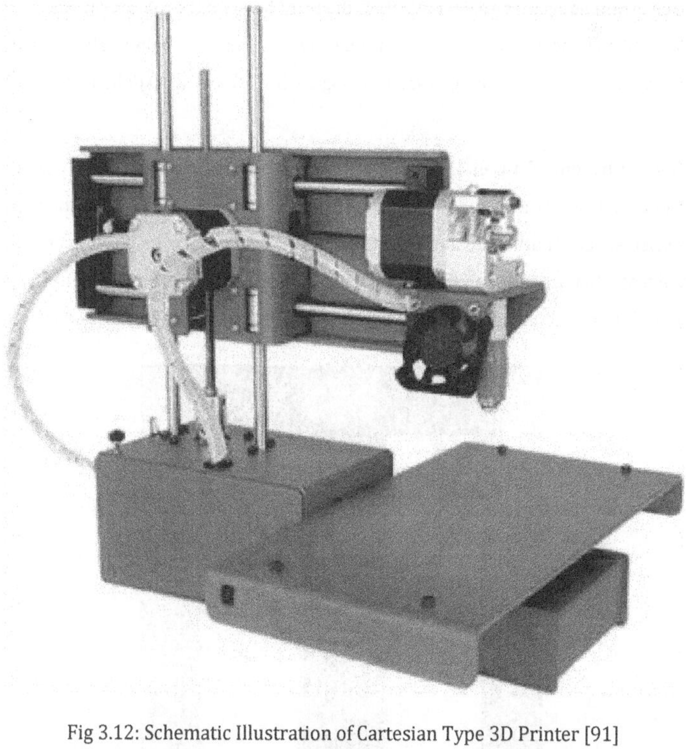

Fig 3.12: Schematic Illustration of Cartesian Type 3D Printer [91]

One of the greatest advantages of Cartesian printers is their popularity. Since Cartesian printers are the most widely used type of printer, there is far more support for users of these printers. In addition, prints from a Cartesian printer tend to have a better surface finish than prints from other printer. This is because Cartesian printers have more rigid axes, which allow less room for error when the print head moves within the 3D space [90].

2. *Delta FDM 3D Printer*

Delta 3D printers are the second most popular type of FDM 3D printer. They can be easily distinguished from a Cartesian 3D printer due to their tall appearance. The unique structure as well is quite intriguing to look at and it's a joy to watch the Delta 3D printer in action. The printhead is connected to the three arms of the printer and they all operate in sync to print the layer of the object. The delta 3D printer works in a triangulation configuration and since the three arms work simultaneously, the build volume of the printer is not a cube but a cylinder and the base is a circle. Instead of positioning the print head according to x, y, and z positions, each of the three arms changes its angle to position the print head on the build platform according to the model. Hitting the precise location is done by a series of trigonometric functions that consider the angle of all positioning arms [92]. Figure 3.13 shows a 3D delta metal printer used in industry.

The biggest benefit of a Delta 3D printer is its speed. The high speeds are possible as the extruder does not carry the weight of the stepper motor and this weight reduction is translated into increased speed. Another advantage of the Delta 3D printer is its height. It is tailor-made for creating objects that are taller and so manufacturers rely on the delta coordinate system when they want to build a 3D printer for manufacturing objects that have a higher Z-axis.

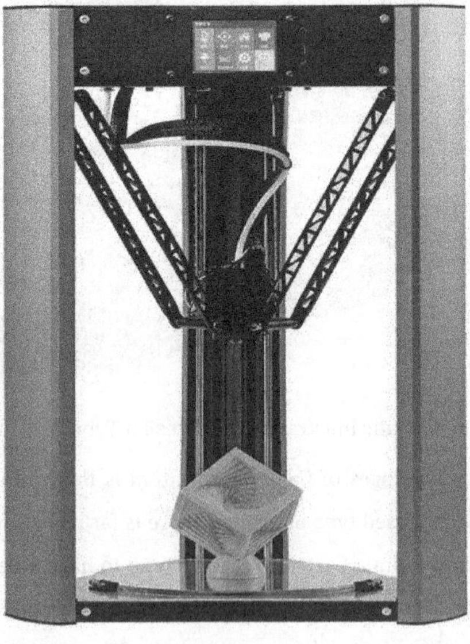

Fig. 3.13: 3D Delta Metal Printer [93]

3. *Polar FDM 3D Printer*

A polar 3D printer looks very different from Cartesian and delta 3D printers, with its rotating circular build plate being the most notable contrast in design. The frame is L-shaped, with the filament spool feeding the extruder from the back similar to most 3D printers with a Cartesian XY (the print head moves in the X and Y axes) mechanical arrangement. Unlike a regular 3D printer, there is no large gantry where the print head moves along the X-axis. Instead, the print head is attached to an extended arm on the upright, parallel to the build platform, and moves up and down [94]. A commercial polar 3D printer is shown in figure 3.14.

Fig. 3.14: Commercial Polar Type FDM 3D Printer [95]

In addition to its rotating motion, the circular print bed also moves forward and backward. In other polar 3D printer designs, the print head is the one the moves forward and backward, with the print bed simply rotating in a stationary position. A polar 3D printer takes up less space than a regular Cartesian 3D printer, although there are also small-form Cartesian 3D printers. But despite its relatively small footprint, it allows for large prints, which is one of its advantages. Although a polar 3D printer certainly looks cool in action due to its rotating print bed and appears less complicated due to having fewer components, its reliability is still unproven. There isn't much evidence concerning its print quality and print speed compared to Cartesian and delta 3D printers.

4. *Robotic Arm FDM 3D Printer*

Robotic arms are most commonly known for assembling components on industrial production lines, especially in large automotive plants as shown in figure 3.15. While 3D printing has begun to incorporate robotic arms into their production process, most notably seen in the 3D printing of homes and buildings, this technology still remains in the development stage. Although not a commonly used printing process, this FDM printing method is beginning to see an increase in use. This is because the process is not fixed to a printing plate, making it much more mobile.

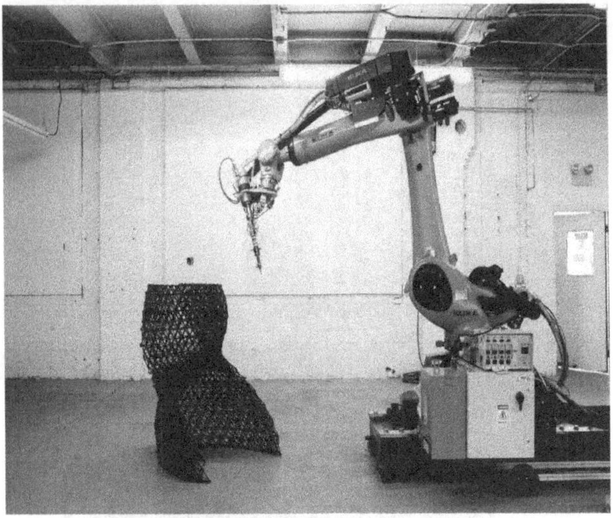

Fig. 3.15: Robotic Arm 3D Printer [96]

In addition, thanks to the flexibility when positioning the FDM 3D printer head, it is easier to create complex structures. It should be noted, however, that the final print quality is not as good as conventional Cartesian printers. With a multi-axis robotic arm, a robotic 3D printer can move in various directions and 3D print at different angles, expanding the freedom of creating complex shapes even further. Additionally, this type of technology typically doesn't use any support structures. Although this means that objects made with robotic 3D printing have to be self-supporting, this challenge can be resolved by reorienting the building platform, making overhangs possible. The combination of robotics and 3D printing is also beneficial in terms of better material management. 3D printing reduces material waste, as the material is deposited only where needed. Robotic 3D printing therefore presents a sustainable approach to manufacturing [97].

3.3.4. Printing Surface

In 3D printing, perfecting the first layer is such an important step that long and detailed guides have been written about it. So many things can go wrong that will hinder you from getting the first layer right, which makes it a good indicator for how well the rest of the print will go. The first layer is also a good time to troubleshoot, as you can anticipate problems that might come up during the rest of the print without wasting too much time and filament material. several factors come into play when trying to get the perfect first layer. The choice of printing bed is one of these factors. The right printing bed should do two things: provide enough adhesion for the filament material to stick and prevent warping. A print that does not stick to the printing bed will stick instead to the extruder, resulting in an uncharacteristic blob of molten filament. Warping occurs when the material at the outer edges of the first layer cool at a faster rate compared to the material in the inner sections, resulting to a deformed base [98]. Figure 3.16 depicts the component printed on the glass bed surface.

Fig. 3.16: Shows a Component Being Printed on a Glass Print Bed [98]

In theory, printing on bare aluminum works well since the high conductivity of aluminum ensures that it is heated uniformly. This means that the first layer of the print also heats up uniformly, reducing the risk of warping. Aluminum is also a relatively sturdy material, which means it can withstand the somewhat rough methods needed to separate the print from the bed. The major drawback of using a print bed made of aluminum is the fact that aluminum expands a lot under high temperature. Since print beds need to be anchored by screws to the frame of the 3D printer to make a stable surface, the thermal expansion of aluminum usually results in the print bed itself getting warped.

Print bed can be made up of different materials such as:

- Glass
- FlexPlate
- BuildTak
- Meta

The general requirements of a print bed are as follows [99]

1. A smooth printing process: If your flat bed is cold, it can cool down your first layer before the next layer is applied. This means inferior binding between layers and compromised model strength.
2. Easy removal of completed print: You want to be able to pull your model from the flat bed easily. If fixed too well to the flat bed, a hard yank could damage your finished print.
3. Easy to clean: Although you don't want your model to stick too much to the flat bed, it is still important it stays in place during the printing process to allow for accurate printing. This often means using substances like glue which then needs to be cleaned for the next print. You want to minimize this process, especially if you are an avid printer.

3.3.5. Electronics

Electronics represents the brain of the 3D printer. It will be responsible for deciding the movements and actions in each moment. Whenever an electronic circuit is used that has to interact with the environment, it will require sensors and actuators.

- A sensor is a device to find physical or chemical quantities and transfer them into the electrical property.
- An actuator is the opposite, from electrical signals generate physical or chemical magnitudes. For example, an electric motor that transforms electricity into motion.
- Microcontroller: It is a programmable integrated circuit, capable of executing the orders recorded in its memory. That is the brain of the printer.
- Heating system: The printers have a resistance as actuators to heat the extruder and a heated bed. Temperature sensors are used to determine the temperature of these components.
- Movement system: As motors are the motors. The minimum will be one for each axis and another motor for the extruder. The movement part uses sensors to know what position it is in. The most used are the end of a career.
- Interface: It is the part of the electronics that allows the user to exchange information with the printer. The most common is an LCD screen, but it can also be through a computer using a USB cable, WIFI or Bluetooth. Figure 3.17 shows various electronics components present in a simple 3D printer setup in FDM.

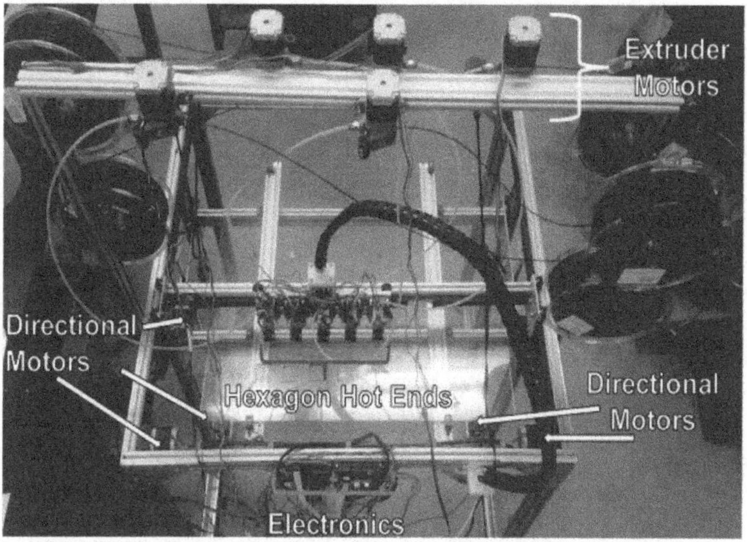

Fig 3.17: Simple 3D Printer Setup with Various Electronic Components [100]

CHAPTER 4

4. Materials Used in Fused Deposition Modeling

4.1. Introduction to Filament Materials

Conceptually, any material which can flow and then harden can be used to create objects by FDM. This includes a wide range of materials from thermoplastic polymers to concrete slurry, cake icing to lab-grown human collagen. This is a rapidly changing field, but current FDM 3D printing is almost entirely focused on extrusion of thermoplastics and thermoplast-fills. Plastics absorb a certain amount of atmospheric moisture, which can flash into steam during plastic melting and cause inconsistent behavior and jamming of the FDM extruder. This means that it is important to keep all printing filament dry in either a dehumidified space or a desiccator cabinet. In general polymers can be split into two categories: thermoplastics and thermosets. Thermoplastics are polymers that, after forming, can be melted and formed again (e.g. polyethylene). Thermosets are polymers that once formed cannot be melted and reformed (e.g. epoxy). Since thermosets degrade instead of melting when heated, they cannot be used in FDM [101]. Figure 4.1 shows pictorial representation of filament materials used for printing.

Fig. 4.1 Filament Materials Used for Printing [102]

4.2. Properties of Polymers for FDM

When choosing a thermoplastic polymer to use in a FDM, there are many properties of polymers that can contribute to how well the part prints and the resulting properties of the printed parts.

The properties of interest for how well a polymer prints include melting temperature (Tm), glass transition temperature (Tg), and the coefficient of thermal expansion (CTE). The melting temperature will affect the temperature the extruder needs to be at to print the polymer. Tg along with the CTE relate to how much thermal stress is developed during printing which can affect bed adhesion. Solvent resistance, strength, and ductility are also important properties. All these properties are a function of the polymer chemistry and microstructure of the polymer. The strength and ductility of the printed part are also directly affected by the chemistry and microstructure as well.

Materials are usually graded along 3 categories: mechanical performance, visual quality, and process. In this case, we further break down these categories to paint a clearer picture of the polymer's properties. The choice of material really depends on what the user wants to print, figure 4.2 shows the key decision criteria needed to choose a printing material.

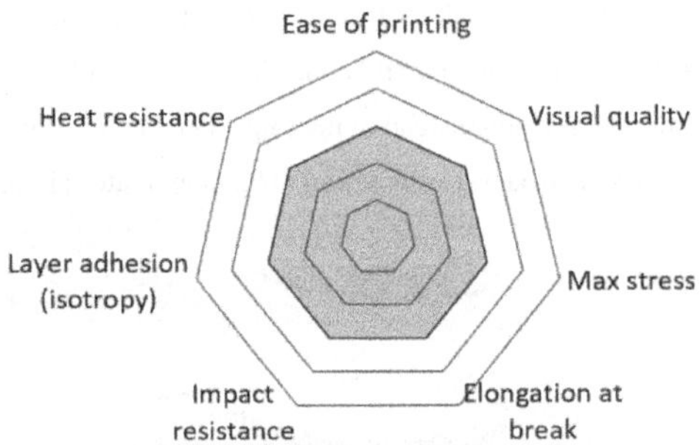

Fig. 4.2: Decision Criteria for Material Selection [103]

- Ease of printing: How easy it is to print a material: bed adhesion, max printing speed, frequency of failed prints, flow accuracy, ease to feed into the printer etc.
- Visual quality: How good the finished object looks.
- Max stress: Maximum stress the object can undergo before breaking when slowly being pulled.
- Elongation at break: Maximum length the object has been stretched before breaking.
- Impact resistance: Energy needed to break an object with a sudden impact.
- Layer adhesion: How good the adhesion between layers of material is, the better the layer adhesion, the more isotropic the object will be.
- Heat resistance: Max temperature the object can sustain before softening and deforming.

4.3. Materials Used in FDM

A wide range of materials are being used in FDM, the following list includes the most common types of filaments used in fused.

4.3.1. Polylactic Acid (PLA)

Polylactic Acid (PLA) is one of the most commonly used 3D printing filaments mostly by enthusiasts, hobbyists and students. It is made from plant-based starches and so it is a bio-degradable material. Moreover it is easy to print with; it is affordable and easy post-process. All these characteristics make PLA the most favoured choice among users to make prototypes. But PLA is not suitable for end-use applications as it does not offer enough mechanical strength. This fact limits its use to prototypes, educational models, gifting, display & props and architecture. But now-a-days it is gaining immense popularity for the manufacture of prosthetics due to its apparent benefits. However, PLA will be less durable, more fragile, and more sensitive to heat than ABS. PLA has a glass transition temperature of 65 °C and a melting temperature of 178 °C. PLA is not the most heat resistant material; this is why it is particularly suitable for decorative objects with no mechanical constraints [104]. Figure 4.3 shows a typical example for PLA filament used for printing.

Fig. 4.3: A 3D Printing Filament [105]

1) PLA Advantages

- Biodegradable
- Low shrinkage
- Easy to post-process
- Easy and fast to 3D print
- Lower printing temperature
- Layers deposited accurately gives a fine surface finish

2) PLA Disadvantages

- Not really durable
- Bad UV light resistance
- Entraps moisture from the air
- Causes nozzle clogging sometimes
- Becomes softer (can be crooked or warped) at around 140°F (60°C)

3) PLA Applications

PLA filaments have a wide spectrum of applications. You can use it in rapid prototyping, for making 3D printed showcase models, or educational purposes. PLA is also commonly used for architectural models, containers, molds, and consumer goods. Figure 4.4 shows a sample printed from PLA filament.

Fig. 4.4: 3D Printed PLA Sample [106]

4.3.2. Acrylonitrile Butadiene Styrene (ABS)

ABS (Acrylonitrile Butadiene Styrene) is part of the thermoplastic polymers family. As its name implies, ABS is created from Acrylonitrile, Butadiene and Styrene polymers. It is a material commonly used in personal or household 3D printing, which is done using primarily FDM 3D printers. This common thermoplastic material is popular because it is easy to use with a desktop 3D printer, but also because it has some great material properties. Figure 4.5 shows an ABS extruder in the process of laying ABS layer by layer on the platform of a 3D FDM printer.

Fig. 4.5: Process of Laying ABS Layer by Layer [107]

ABS is lightweight and has good impact strength; it is abrasion resistant and affordable. Moreover, ABS polymers withstand a lot of chemical formulas. The glass transition temperature of ABS plastic is 105°C (221°F), making it ideal for use in relatively safe machines that are easy to operate (the safety of household machines is important). ABS material has a low melting point, which makes it a material that is not really adapted to high heat situations. However, this low melting point makes ABS easy to machine with desktop 3D printers, or during an injection molding process.

1) ABS Advantages

- Recyclable
- Abrasive resistant
- Can be machined
- Stiffness and durability
- Easy to post-process and paint
- Many colors and composites available
- Withstands higher temperatures 212°F (around 100°C)

2) ABS Disadvantages

- UV sensitive
- Harder to print with
- Smells bad when heated
- Higher printing temperature

3) ABS Applications

This plastic material is for example used to create toys. But it can also be used to print enclosures for electrical or electronic assemblies, sports equipment, parts for the automotive industry, or the medical sectors. Indeed, this is an ideal material to manufacture low-cost prototypes and architectural models for engineers or research departments, as well as to create low-cost cost medical prostheses or material handling equipment. Many plastic end-products can be manufactured using ABS material. Figure 4.6 shows a 3D printed component made of ABS filament in a FDM 3D printing machine.

Fig. 4.6: Component Printed from ABS Filament in a FDM Process [108]

4.3.3. Polycarbonates (PC)

Polycarbonates (PC) are among the most widely used industrial thermoplastics owing to the material's excellent impact strength and temperature resistance. The mechanical properties of PC make this material ideal for demanding engineering environments or applications requiring high flexural strength and tensile strength [109]. PC is a thermoplastic polymer containing carbonate groups in its structure that makes the filament much stronger and a nice alternative to ABS. Figure 4.7 shows a custom assembly pallet from PC filament in FDM process.

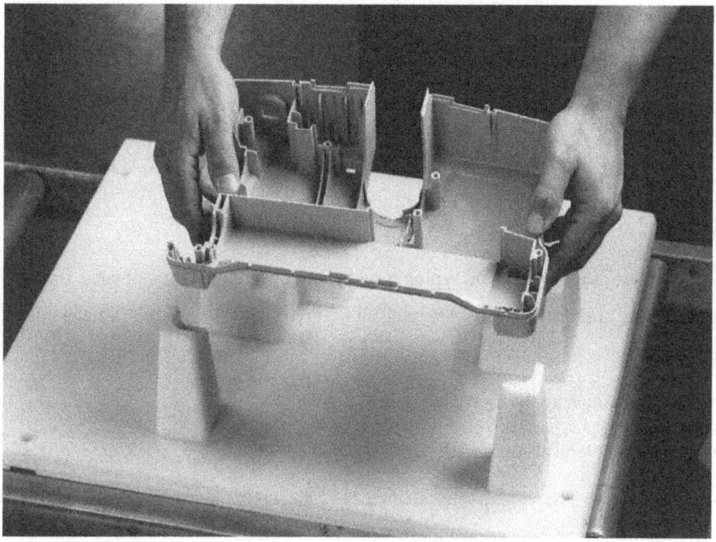

Fig. 4.7: Custom Assembly Pallet Made from PC [111]

Other advantages of 3D printing in PC include extremely high resistance to impact and the combination of a great flexural and tensile strength. As a result, Polycarbonates can be bent at room temperature without breaking. The color of PC materials is usually transparent allowing it to be used in parts where it is necessary to visualize the flow of liquids. Besides, the material has high electrical insulating characteristics and is lighter and less dense than ABS. Accuracy, stability and the ability to withstand functional testing are other traits to be expected from parts 3D printed in PC. Polycarbonates are being used in automotive, aerospace and medical applications for both functional prototypes and end parts [110].

1) PC Advantages

- Strong
- Durable
- Heat resistant
- Can be sterilized
- Can be transparent
- Easy to post-process
- Can be flame retardant
- Easier to print with compared to ABS

2) *PC Disadvantages*

- UV sensitive
- Absorbs moisture a lot
- Can be hard to print with
- High printing temperature

3) *PC Application*

Polycarbonate is useful in prototyping applications where sheet metal lacks transparency, non-conductivity and insulation. Molds made of PC are regularly used of urethane and silicone casting [112].

Some of the notable applications are:

- Medical - Medical equipment parts.
- Automotive - Headlamp lenses, inner lenses, clusters, door handles, finishing panels, radiator grilles, hubcaps, spoilers.
- Electric and Electronic Parts - Switching relays, connectors, sensor parts, deflecting yokes, disc carriers, LCD parts, backlit panel parts.
- Machinery - Electric tools, vending machine parts, fibre bobbins.
- Safety - Safety goggles, helmets, smoke detectors, revolving lights, etc.
- Precision Parts - Camera parts, watch parts.
- Office Automation Equipment - Lap-top personal computer housings, PDA housings, CD-ROM drive parts, laser printer chassis, OA equipment housings, digital camera housings.

4.3.4. Polyethylene Terephthalate (PET)

Today, Polyethylene Terephthalate (PET) is a popular 3D material. Another common use for PET is in everyday plastic bottles. This plastic is both stable and harmless, emits no unpleasant or harmful odors, and is 100% recyclable. In its raw state, the filament has no color and is crystal clear. Once exposed to cold or heat the material quickly changes to a non-transparent state. This material is the second alternative to ABS. PET does not require a heated bed. This material has a glossy finish, is flexible, and is food safe, which makes it a popular choice for many consumer products [113]. Figure 4.8 shows a lid made from PET foil oh FDM mold.

Fig. 4.8: Lid Made from PET on FDM Mold [114]

1) PET Advantages

- Recyclable
- Can be food-safe
- Moisture resistant
- Has some flexibility
- Can be semi-transparent
- Layers look nice and smooth
- Quite tough and shock-resistant
- Can be post-processed and painted
- Printing temperature ranges between PLA and ABS

2) PET Disadvantages

- Medium heat resistance
- Changes color while printing
- Becomes softer at around 158°F (70°C);
- Becomes brittle if overheated during printing

3) PET Applications

- Water proof applications
- Snap fit components
- Planter Pot

4.3.5. PETG

PET has become popular as an FDM 3D printing material in the modified version called PETG. It is a modified version of PET, with the "G" standing for "Glycol," a molecule added during the polymerization process to make it less fragile, more durable and easier to use. This version is preferred for FDM filaments used in additive manufacturing and is generally translucent but also available in a range of colors. PETG exhibits excellent mechanical properties and also boasts resistance to moisture, chemicals and heat, particularly when compared to PLA. Figure 4.9 shows PETG filament and 3D printed vacuum nozzle. It is superior to ABS, with greater impact resistance and durability. Thanks to a strong layer adhesion, it can better withstand UV light and is generally less challenging to print with. Additionally, PETG emits no smelly fumes. When printing with PETG, it's less likely that your part will warp or shrink because of temperature changes. The greatest limitation of PETG is it absorbs moisture from the air.

Fig. 4.9: PETG Filament and 3D Printed Vacuum Nozzle [115]

Air humidity negatively affects the material and can lead to failed prints. For this reason, the filament should be stored in a dry environment [116].

1) PETG Advantages

- Recyclable
- Prints are strong
- Moisture resistant

- Has low shrinkage
- Layers look nice and smooth
- Good heat and stress resistance
- Can be food-safe and even sterilized
- Printing temperature is between PLA and ABS

2) PETG Disadvantages

- Sensitive to UV light
- Scratches appear easier
- Can change color while printing
- Becomes softer at 176°F (around 80°C)
- Less PETG variations compared to PLA and ABS

3) PETG Applications

When working on new concepts, designers and engineers can take advantage of its transparency, moisture and chemical resistance to make great functional prototypes such as containers for liquids, signage and graphic displays, enclosures for electrical equipment, etc. Figure 4.10 shows a 3D print with PETG.

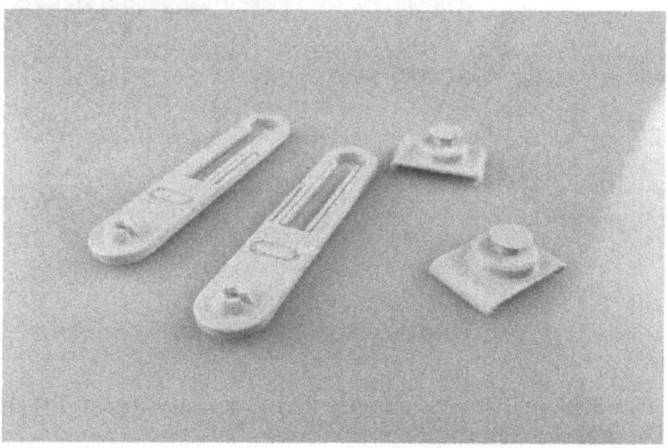

Fig. 4.10: 3D Print with PETG [117]

The most obvious use of PETG 3D printing is for the design and prototyping of the packaging for consumer products that will be later made out of some form of PET when it enters mass production. Additionally, because PETG remains a durable and performant polymer, engineers and machinists can create true end-use parts including snap fits and living hinges, protective guards or custom machine parts.

4.3.6. *Thermoplastic Polyurethane (TPU)*

Thermoplastic polyurethane, or TPU, is a very common form of elastic polymer, or elastomer, capable of being used on any properly equipped FDM 3D printer. Of the many kinds of flexible 3D printer filaments, TPU is slightly more rigid, making it easier to extrude and use. It is a medium-strength material with very high flexibility and durability, able to withstand much higher compressive and tensile forces than its more common counterparts PLA and ABS [118]. When printing, TPU exhibits minimal warping and shrinkage and, like PLA, doesn't require a heated bed, although it is recommended. The ideal print temperature is between 210 and 230 °C, but this may vary from manufacturer to another. Figure 4.11shows a thermoplastic polyurethane part produced with FDM.

Fig. 4.11: Thermoplastic Polyurethane Part Manufactured with FDM [119]

TPU doesn't produce any notable levels of fumes while printing, but it's not considered to be food safe. And though it's a non-soluble material, it is hygroscopic, meaning that it will slowly absorb moisture from its surroundings, degrading over time. One downside to TPU is the printing speeds, which are rarely higher than 30 mm/s. This is because it's difficult to push the proper amount of filament through the hot end, resulting in prints taking a lot longer to complete.

1) *TPU Advantages*

- Flexible
- Low shrinkage
- High strength
- Wear resistant
- Good abrasion resistance
- Can be semi-transparent
- Great resistance to oils and grease
- Withstands temperatures of up to 176° F (80°C)

2) *TPU Disadvantages*

- Difficult to glue parts
- Difficult to post-process
- High printing temperature
- Fewer variations than some plastic filaments.

3) *TPU Applications*

TPU filament is commonly used to 3D print a wide variety of items. For everyday use, TPU is great for making custom phone cases, rubber mats, and stress toys. In the RC area, TPU can be used for tires or shock absorbers and is even occasionally utilized to make O-rings and seals. TPU is fantastic for flexible parts or casings as well as parts that need to have resistance to sudden impacts and movement. The automotive industry has turned to TPU material and implemented it into the manufacturing process of vehicles [120]. In recent years the use has risen considerably. Consequently it can often be found in instrument panels, sensors and levers as well as many other areas. The aerospace industry is also taking the use of TPU 3D printing seriously as it too has increased its use, understanding exactly what it has to offer.

Some common applications include:

- Seals, gaskets
- Surface protection covers
- Mats, liners
- Bellows
- Wheels, rollers
- Hoses. Tubes
- Ducts
- Manufacturing aids

4.3.7. *Acrylonitrile Styrene Acrylate (ASA)*

ASA – is Acrylonitrile Styrene Acrylate and it is a thermoplastic originally created as an alternative to ABS. The ASA is an amorphous thermoplastic terpolymer material similar to ABS. The difference between these two materials at the structural level is that the ASA uses an acrylic elastomer and the ABS an elastomer of butadiene. The ASA is referred to as engineering plastics because it maintains its appearance and resistance to impact even after being exposed to long periods of time in the open air, rain, cold and salt water of the sea. ASA is slightly superior to ABS in both mechanical performance and aesthetic quality. That is why it's a material used in many products that we see and use in our daily life: parts of houses (roof coverings), electrical installations (junction boxes), auto parts (front grilles), toys, structural elements for outdoors (pipes and construction profiles) and many more. ASA also has the best bridging properties of any FDM material. This means that you can print parts with a sparser interior fill than with ABS or other FDM plastics, which enables greater material savings for applications where cost and/or weight are critical [121]. Figure 4.12 shows a UV resistant ASA thermoplastic.

Fig. 4.12: UV Resistant Component Printed from ASA Thermoplastic [122]

1) *ASA Advantages*

- Strong and durable
- High UV-light and weather resistance
- Withstands high and low temperatures
- Lower color changes throughout the time
- More resistant to stress cracking (even more than ABS)

2) *ASA Disadvantages*

- ASA is hygroscopic, which means it absorbs water from the air
- May be tough to work with
- May produce toxic smoke when burnt
- Some grades and filaments cost much
- ASA melts with some other plastics giving rise to moldings of poor strength
- Can't withstand some concentrated acids, hydrocarbons, esters, ethers and ketones

3) *ASA Applications*

- Automotive
- Medical products
- Consumer Products
- Electronics
- IT and Communication
- Outdoor Products
- Building and Construction
- Sports equipment
- Exterior signage

4.3.8. ULTEM

Ultem is the ideal thermoplastic choice for demanding, high-heat applications. This polyetherimide material can withstand continuous operating temperatures of 340° F while maintaining strength and rigidity. It is easily fabricated and bonded and offers excellent chemical resistance and high dielectric strength. High thermal properties come with inherent flame resistance, including a low smoke evolution and low smoke toxicity. It is known to be a material that helps to achieve a significant level of stability including long-term hydrolytic stability and dimensional stability. This includes low creep sensitivity and low, uniform coefficient of thermal expansion, while also making it possible to create highly reproducible part-to-part dimensions [123]. Figure 4.13 shows an Ultem 9085 aerospace filament used for 3D printing in FDM processes.

Fig. 4.13: Ultem 9085 Filament Used for 3D Printing in FDM [124]

Even when ULTEM is exposed to high temperatures, it continues to behave in a way that delivers an exceptional level of strength and modulus. ULTEM is used in the automotive and aerospace industry due to the fact the material can withstand a lot of chemicals such as automotive fluids, fully halogenated hydrocarbons, alcohols and aqueous solutions. Over a wide range of temperatures and frequencies, ULTEM material also delivers a stable dielectric constant and dissipation.

1) Ultem Advantages

- Heat resistant
- Stress resistant
- Chemical resistant
- Strong and dielectric
- Some types are biocompatible and pass toxicity/smoke tests

2) Ultem Disadvantages

- Can be hard to 3D print
- High printing temperature
- Produces fumes 302 – 392°F (150 – 200°C)

3) *Ultem Applications*

ULTEM material is used in the aerospace industry because of the impressive balance of flame retardancy, low smoke toxicity and low smoke emissions. Therefore, the material can be found in applications which include personal service units, oxygen panels, ventilation system components, connectors, cable ducts, latches, door handles and interior cladding parts. Figure 4.14 shows a real time component printed from Ultem filament in FDM processes.

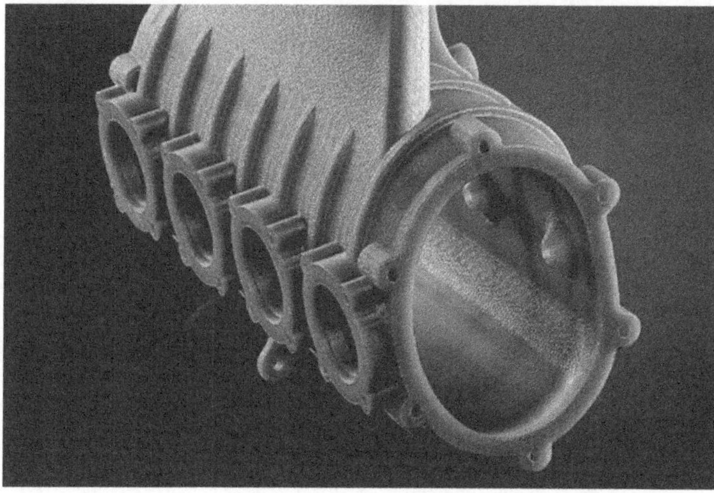

Fig. 4.14: Automotive Component Made from Ultem Filament [125]

Automotive manufactures require a high level of chemical resistance, ULTEM offers this. It is thermally stable and is known to be a cost effective alternative to metal where in some instances it has the ability to replace steel while also being light enough to be used in place of aluminum. It can also be found in transmission and ignition components, sensors, thermostat housings and throttle bodies.

Due to the high demanding requirements, ULTEM material is very well suited for electrical applications. Because of the ability to deal with the high demands, it will be found in connectors, MCB components such as shafts, levers and housings, the internals of hard disk drives, BiTS, PCB's, MCCB internals, FOUP's, Plenum Devices, LCD projector internals and fuel cell components. With its great features ULTEM has been certified for use in food production products. Due to its resistance to heat, it can be found in the production of baking sheets, patterns, dies and fixtures.

CHAPTER 5

5. Research Trends in FDM of Thermoplastic Polymers - ABS

5.1. Introduction to Thermoplastic Polymers

It is agreeable that 3D printing technology is one of the most significant modern innovations that allow the manufacturing process of any component at a very short period and simpler steps. Also, the FDM process allows the parts with complex geometries to be designed and produced more straightforwardly compared to the conventional methods. When a 3D printer operates in an optimum condition, it is assumed to be printing the component at its best quality. FDM is a slow printing process, which is limited to materials with a low melting point. Furthermore, the print dimensions are restricted, and it gives a rough surface finish to the printed part. It also brings about many advantages over these drawbacks, such as high reliability, a wide range of low-cost filament material availability and low maintenance and initial investment cost. The FDM method can be utilized to print polymers, polymer matrix composites (PMC), bio-composites or polymer ceramic composites (PCC), nano composites and fibre-reinforced composites (FRC). ABS (acrylonitrile butadiene styrene), PLA (polylactic acid), Nylon/ polyamide, ASA (acrylonitrile styrene acrylate), PET (polyethylene terephthalate), PETG (polyethylene terephthalate glycol-modified) and PC (polycarbonate) are the most commonly used polymers, while PEEK (polyether ether ketone), PEKK (polyetherketoneketone) and ULTEM (polyetherimide) are known as high performance polymers, and TPE (thermoplastic elastomers) is the newest flexible polymer material used in FDM. A simplified process flow chart of FDM is shown in figure 5.1. The thermoplastic polymer matrix plays a major role in FDM fibre-reinforced composites. Both matrix and fibres have a great influence on the mechanical properties and the defects of the composite. ABS, PLA and nylon are the most common types of filaments used in FDM because of the low melting temperature. Due to low strength and functional properties in pure polymer printed parts, these are mainly used to build concept samples or prototype samples.

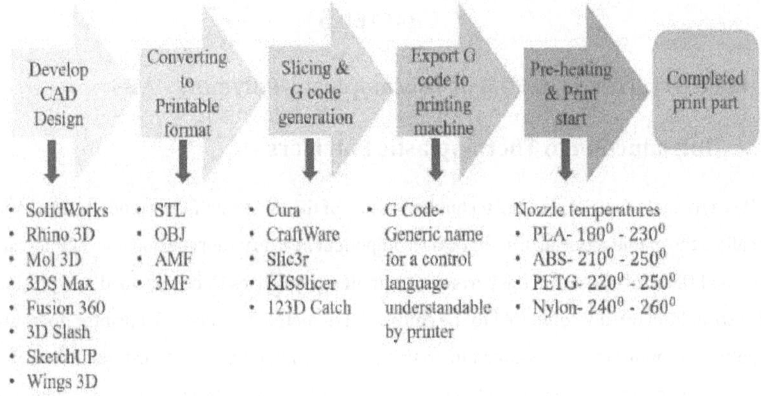

Fig. 5.1: Simplified Process Flow Chart of FDM Including different Types of Software's Used [126]

5.2. Acrylonitrile Butadiene Styrene Polymer in FDM

Acrylonitrile–butadiene–styrene (ABS) is a popular engineering thermoplastic and it is the most common material used in fused deposition modeling (FDM) technology. This technology is nowadays used for the production of prototypes and functional parts as well. Acrylonitrile–butadiene–styrene (ABS), because of its unique properties, such as an excellent mechanical response, chemical resistance, fine surface finish, and good processing characteristics, is widely used in various industries. In order to predict the mechanical behavior of FDM parts, it is necessary to understand the material properties of the raw FDM process material, and the effect that FDM build parameters have on material properties which are anisotropic in the case of additive manufactured ABS parts. The process control parameters that were likely to affect the properties of ABS - FDM parts are listed below.

5.2.1. Raster Width

Raster width is the thickness of the bead (or road) that the FDM nozzle deposits. It can vary from 0.3 mm (0.012 inch) to 1 mm (0.0396 inch) in general for various machines and these values are given by the manufacturers. The figure 5.2 shown below shows the raster width, Air gap and the raster orientation.

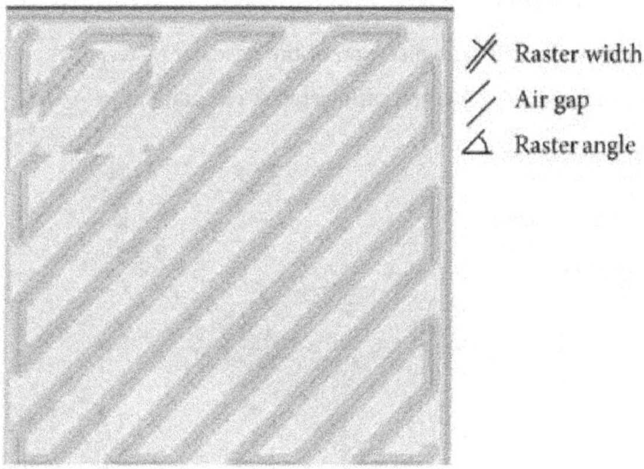

Fig 5.2: Illustrating Raster Width, Air Gap and Raster Angle [127]

5.2.2. *Air Gap*

Air gap is the space between the beads of FDM material. The default is zero, meaning that the beads just touch. It can be modified to leave a positive gap, which means that the beads of material do not touch. The positive gap results in a loosely packed structure that builds rapidly. The air gap value can also be modified to leave a negative gap, meaning that two beads partially occupy the same space. This results in a dense structure, which requires a longer build time. Figure 5.3 illustrates the schematic representation of air gap.

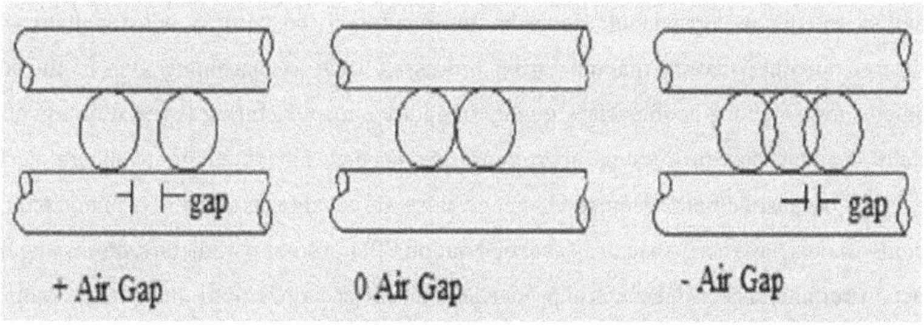

Fig. 5.3: Schematic Representation of Various Types of Air Gaps [128]

5.2.3. *Model Build Temperature*

This parameter refers to the temperature of the heating element for the model material. It controls the viscosity of molten material that is extruded from the nozzle.

5.2.4. *Raster Orientation*

Raster orientation refers to the direction of the beads of material (roads) relative to the loading of the part. Figure 5.4 shows the various raster orientations in FDM.

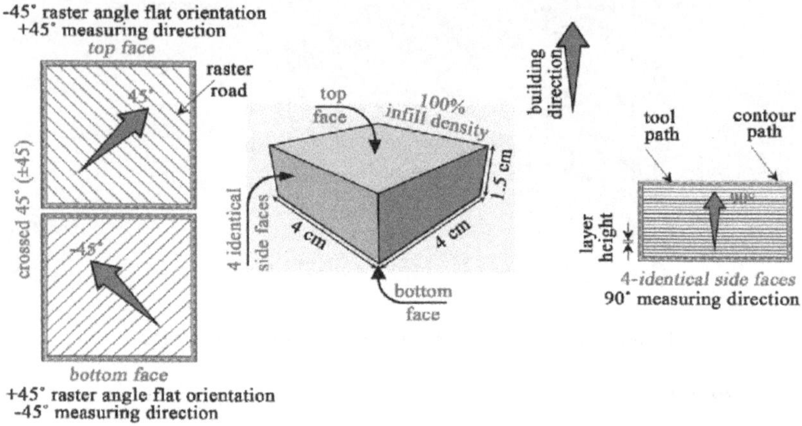

Fig. 5.4: Various Raster Orientations in Fused Deposition Modeling [129]

5.3. FDM Printed ABS Parts

Although FDM processes have significant industrial value for manufacturing complex components, there is a need to carry out proper research focused on prominent aspects such as surface roughness. In additive manufacturing, in general, pre-processing and post-processing activities should be carried out. However, the quality of the parts is not adequate when compared to other mature manufacturing processes, such as machining. One of the main problems for obtaining good surface quality in additive manufacturing is the staircase effect. Usually manual post-processing operations are needed for obtaining adequate surface roughness because complex geometries compromise the advantages of additive manufacturing. Various studies have been specifically carried out on FDM process parameters, discussing their effect on outputs, such as mechanical properties, and surface topography and quality. Campbell et al. [130] investigated surface roughness for different materials. The authors found that, in the case of ABS material, when using layer thickness of 0.253 mm, the surface roughness values for FDM processes ranged between 9 μm and 40 μm. The optimization of the process parameters used for printing is an adequate strategy for improving part performance in terms of surface finish. Figure 5.5 shows the surface finish of a FDM manufactured ABS component with rough surface finish.

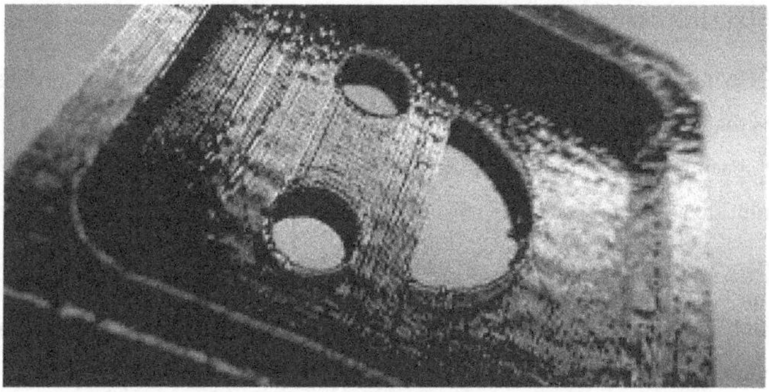

Fig. 5.5: Surface Finish of a FDM Printed ABS Component [131]

Apart from these there are several other parameters which do not have a great influence over the properties and are discussed as follows. Envelope temperature (the temperature of the air around the part), slice height (which is similar to bead width in the vertical direction), and nozzle diameter (the width of the hole through which the material extrudes). These parameters seemed to be either duplicates of the selected parameters, or did not seem to have a relevant connection to the final material properties. The envelope temperature appeared to have an insignificant effect on past parts prototyped with the FDM machine. Furthermore, the temperature in the envelope tends to fluctuate several degree Celsius during a build, making precise control of this variable difficult. The temperature may also vary in different locations in the build volume, increasing the control difficulty. Slice height is the thickness of each layer measured in the vertical or Z direction. Varying the slice height would most likely have the same effect as varying the bead width of the ABS plastic. Since an individual bead of the material is homogenous, varying the geometry of the bead in one dimension should have the same effect as modifying it in the other dimension, causing a redundancy between this parameter and bead width.

One factor that could affect this assumption is the fact that material is likely to have cooled more thoroughly between layers than between roads, potentially affecting the fusion of the plastic. Similarly, the nozzle diameter was assumed to primarily affect the geometry of the bead. Since bead geometry was already being altered via road width (which is controlled by the flow rate of material through the nozzle), altering the nozzle diameter seemed to duplicate the testing. Effecting a nozzle change, which requires a physical transfer and re-calibration for every change, is much more time consuming than changing road width, which is easily modified in software.

5.4. Injection Molds Vs FDM Printed ABS Part

Sung - Hoon et al in their experimental study compared the various properties and the parameters that influence the properties of FDM printed ABS part with Injection moulded parts. A dozen of specimen was tested and the test results showed that air gap and raster orientation have a significant effect on the tensile strength of the given specimen [132].

5.4.1. Tensile Test

Each FDM specimen consisted of 12 layers with various raster orientations. For example, the axial specimen had 12 layers in the zero (loading) direction [00], and the criss-cross specimen had six repetitions of a 45^0 layer followed by a -45^0 layer [$45^0/-45^0$]. Default FDM parts were made with a crisscross raster in which the orientation of the beads alternates from $+45^0$ to -45^0 from layer to layer. Some criss-cross raster specimens were built and tested in addition to the main factorial experiment. Figure 5.6 shows tensile strength of specimen with zero raster angle in comparison with ABS.

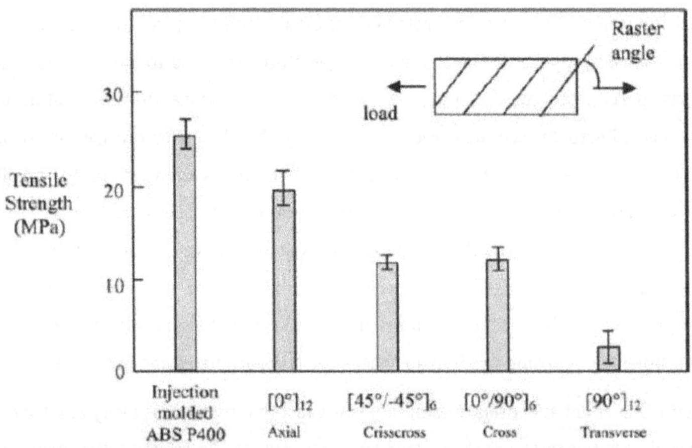

Fig. 5.6: Tensile Strength of FDM with Injection Moulded ABS Part with Zero Air Gaps [132]

The injection molded ABS failed at 26MPa, and the four FDM specimens failed between 10 and 73 percent of the injection molded ABS strength, depending on raster orientation. It is also observed that negative (-0.003) air gap, which makes the specimen more dense and strong. Although the strength of the axial specimens was not increased much with a negative air gap, other FDM specimens exhibited a large increase in tensile strength. The effect of negative air gap is illustrated in the figure 5.7 given below.

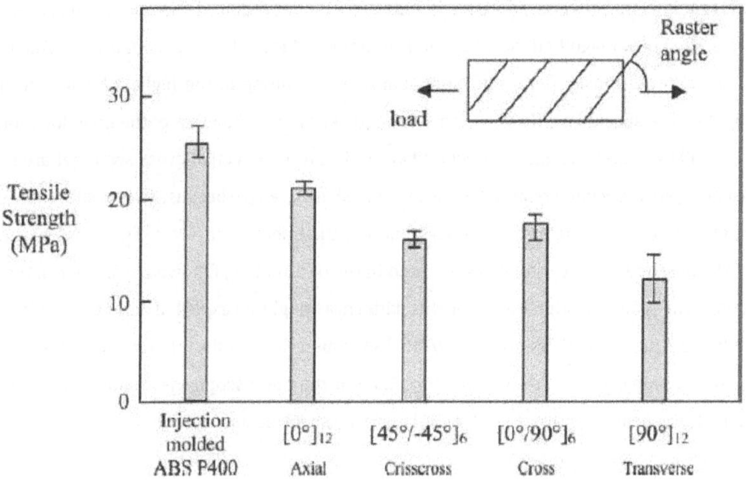

Fig. 5.7: Tensile Strength of FDM with Injection Moulded ABS Part with Negative Air Gap [132]

The [45⁰/-45⁰] raster orientation is of particular interest as the Quick slice software defaults to this raster. This orientation could be looked at as a [0⁰/90⁰] orientation if the part were rotated 45⁰. For these two general cases, the strength ranged between 65 and 72 percent of the injection molded ABS. The failure modes of these specimens are shown in Figure 5.8. All specimens failed in transverse direction except the criss-cross specimen, which failed along the 45⁰ line. The failure modes for specimens with zero air gap were identical to those with -0.003 air gap.

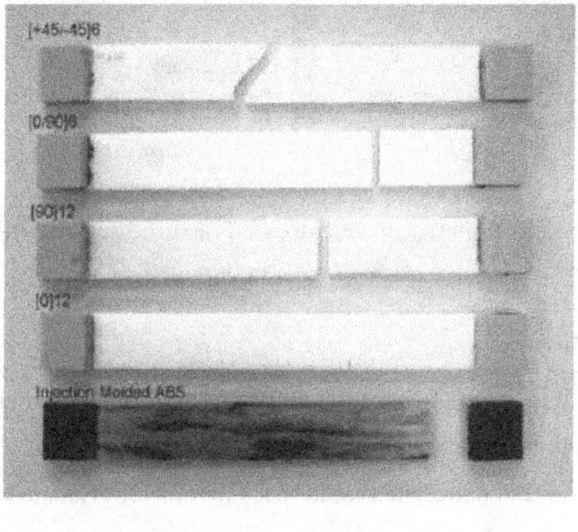

Fig. 5.8: Failure Mode of Specimens with Various Raster Orientations [132]

The relationship between failure loads and failure modes required microscopic observations. Figure 5.10 shows magnified views of the fractured surfaces of the specimens. The Axial [0^0] specimen showed tensile failure of individual fibers resulting in the highest tensile strength among the FDM specimens. However, this strength was lower than that of the injection molded ABS partially because the gaps between fibers reduced the effective cross sectional area. The Transverse [90^0] specimen resulted in the lowest tensile strength because the tensile loads were taken only by the bonding between fibers, and not the fibers themselves. The Cross specimen [0^0/90^0] consisted of a layer of fibers oriented in the 0^0 direction, followed by a layer in the 90^0 direction. The resulting failure load for this pattern, as might be expected, fell between the [0^0] and [90^0] specimen. The Criss-cross [45^0/-45^0] specimen showed shear failure along the 45^0 line in the macroscopic view as shown in figure 5.9, but the microscopic view shown in figure 5.9 revealed the repeated failures of individual fibers by shearing and tension.

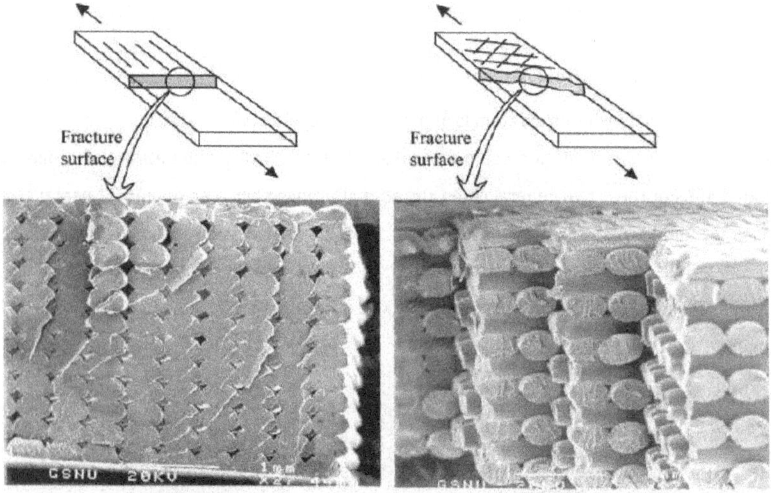

Fig. 5.9: Scanning Electron Microscope Pictures of Fractured Surface with Negative Air Gap [-.003inch] [132]

5.4.2. Shear Test

Daiyan H et al in their experiments studied the effect of shear. The shear specimen results were derived from ten tests for each printer/raster orientation and averaged to give a mean value as well as a 95% confidence interval for each parameter evaluated [133]. The results of these tests are displayed in the Table 5.1.

Table 5.1: Shear Modulus, Yield Strength, and Ultimate Strength Average Values with 95% CIs for the ABS [133]

Orientation	Shear Modulus (MPa)	Yeild Strength (MPa)	Ultimate Strength (MPa)
[+45/-45] flat	740 ± 30	19.1 ± 0.5	28.8 ± 0.2
[0/90]	770 ± 40	21.5 ± 2.0	29.1 ± 0.3
[+45/-45] on edge	610 ± 30	16.0 ± 1.1	23.9 ± 0.7
[0/90] on edge	580 ± 20	14.5 ± 1.3	19.9 ± 1.8
[+45/-45] up right	670 ± 3 0	15.5 ± 1.3	25.8 ± 0.3
[0/90] up right	680 ± 40	15.4 ± 1.3	24.4 ± 0.5

During shear testing only the shear modulus, yield strength, and ultimate strength were calculated as some specimens did not fail before the destruction of the DIC speckle pattern, making strain to failure calculations for some samples impossible. The PC specimens were more apt to fail during testing while the ABS specimens were ductile and very few ABS specimens failed completely. Unlike the results seen in tensile testing, there were several cases of anisotropy seen during shear testing in the ABS specimens. Despite the Young's modulus appearing isotropic in nature during tensile testing, differences of up to 25% were found when comparing shear moduli. There was no significant effect on the shear modulus when varying raster orientation; however, when print orientation was varied the flat orientation was found to have a 25% higher modulus than the on-edge samples and 12% higher modulus than the up-right samples. The same results were seen when evaluating yield strength as raster orientation did not appear to have an effect on the sample strength. When evaluating printer orientation the flat orientations featured the highest values and the on-edge and up-right specimens being approximately equal in value. Finally, assessing the ultimate strengths revealed that the [0/90] on-edge specimens had 32% lower values than the [0/90] flat specimens. This dissimilarity appeared to be due to weak adhesion between the deposited layers of material as the [0/90] on-edge specimens appeared to delaminate along the notched area while most specimens had shear flow along the notched region. The ultimate strength of the up-right specimens was also lower than the flat specimens; however, no delamination occurred and the maximum difference was halved to 16%. Images illustrating the scatter seen in the shear stress-strain curves and the average stress-strain behavior of ten [+45/-45] flat orientation ABS samples are shown in Figure 5.10.

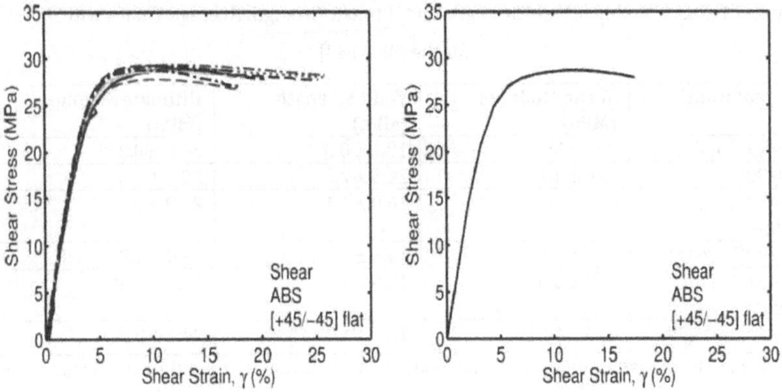

Fig. 5.10: Average Stress-strain Behavior of Ten [+45/-45] Flat Orientation ABS Samples [134]

The fracture surfaces of the shear specimens were also evaluated. Due to the ductile behavior of the ABS specimens very few specimens were brought to complete failure as destruction of the DIC speckle pattern occurred well before specimen failure. Therefore, the ABS specimens fracture surfaces were not evaluated.

5.4.3. Compression Test

For compression test the complexity of the test increases with more parameters hence, build direction was taken as the only examined parameter and the other possible parameters are not considered. "Axial" (horizontal) and "Transverse" (vertical) directions were defined as shown in Figure 5.11. With a fixed raster angle of [45^0/-45^0] and zero air gap, the effect of build direction on compression strength could be measured.

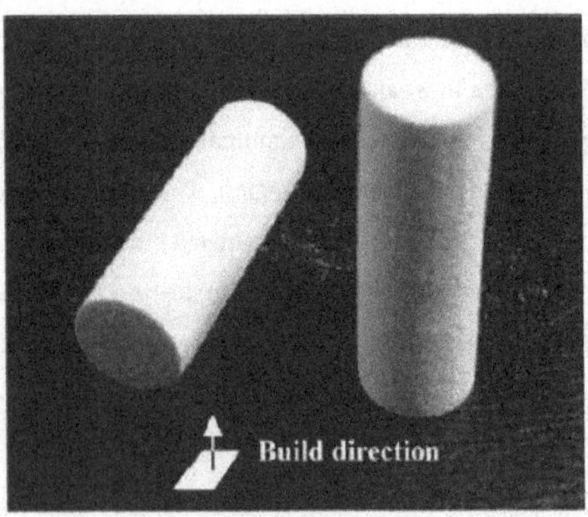

Fig. 5.11: Compression Specimens with Two Different Build Directions [132]

Figure 5.12 shows higher compressive strengths compared to the tensile strengths shown in Figure 5.7. Higher compressive strengths are often observed in polymers, and for bulk ABS materials. For example, tensile strengths of typical bulk ABS range from 32 to 45MPa while their compressive strengths range from 65 to 90MPa (ASM, 1988). The Transverse specimen had 15 percent lower compressive strength than the axial specimen. In general, the compressive strengths of FDM specimens ranged from 80 to 90 percent of those for injection molded ABS.

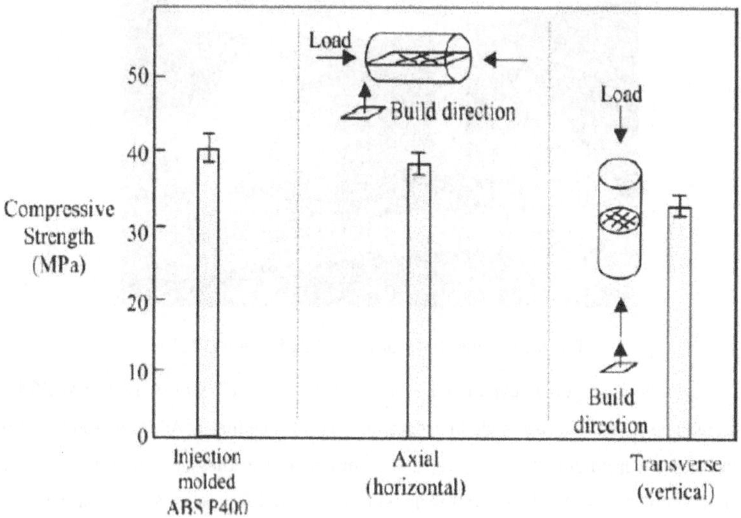

Fig. 5.12: Compressive Strength of Specimens with Different Build Direction Compared with Injection Molded ABS [132]

5.4.4. *Flexural Test*

Flexural properties are largely specified properties for 3D printed products. Flexural strength is the ability of a material to withstand bending forces applied perpendicular to its longitudinal axis. The stress induced due to flexural load is combination of compressive and tensile load. The flexural properties are calculated in terms of maximum stress and strain that occurs at outside surface of test bar. And Flexural modulus is a measure of the stiffness during the first or initial portion bending process. Figure 5.13 depicts the 3-point flexural loading test. A bar of rectangular cross section rests on two supports and is loaded by means of a loading nose midway between the supports.

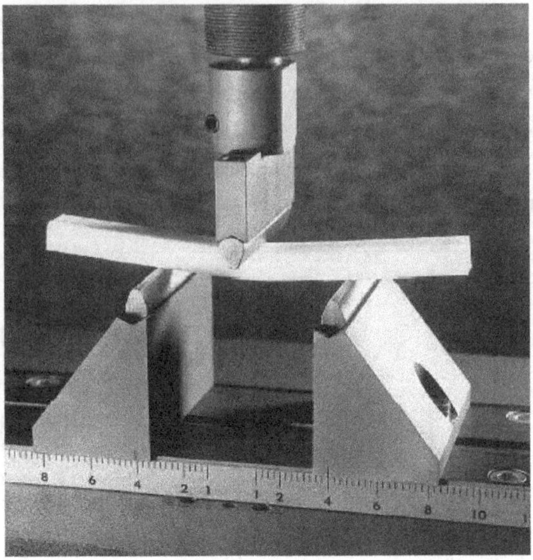

Fig. 5.13: 3-point Flexural Loading System [135]

Divyathej M V et al conducted Flexural tests with 127mm X 12.7 mm X6.4 mm specimens. The specimens are made in accordance with Procedure of ASTM methods D790.[5] The specimens were divided into 4 groups, of 0.1mm layer thickness, 0.15mm layer thickness, 0.2mm layer thickness [135]. Flexural tests were conducted following ASTM standard D 790 Universal testing machine having flexural test fixture such as specimen support and loading nose is used. Machine should be fitted with a load cell and internal extensometer for recording the load and deflection of specimen at any point accurately. The surface of specimen support and loading nose had a surface radius of approximately 3.2 mm. Specimen was placed on support centrally with load axis perpendicular to the 1loading nose. The test was initiated by applying the load to the specimen at the specified crosshead rate. The deflection was measured either by a gauge under the specimen in contact with it in the center of the support span. Flexural strength Fs is calculated from the peak loads in the flexural tests, using the following formula given in equation 5.1.

$$F_s = \frac{3PL}{2bd^2} \tag{5.1}$$

Where P is the peak load in N, L is the distance of sup-ports in mm, b is width of specimen in mm and d is thickness of test specimen in mm. the results obtained from the flexural test are tabulated as shown below in Table 5.2.

Table 5.2: Flexural Test Results [135]

Parameters	Unit	Layer thickness			Injection moulded
		0.1mm	0.15 mm	0.2 mm	
Peak load	Kg	17.8	17.7	18.8	19.9
Deflection	Mm	14.5	14.5	14.9	14.2
Flexural stress	MPa	51.079	51.151	51.505	57.937
Deflection	%	13.81	13.81	14.19	14.06
Flexural Modulus	MPa	804.767	811.698	8.3.052	778.042

When loaded in the z direction, flexural strength obtained is maximum for injection moulded samples compared to FDM samples and it is equal to 57.937 MPa. The molecular orientation in the specimen has a significant influence on the test results. The sample having a high degree of molecular orientation perpendicular to the applied load will show higher flexural values. In the injection moulded samples molecular orientations as well as bond are higher compared to FDM samples. The strain rate, which depends upon testing speed and specimen thickness can affect the results. The results of FDM samples with 0.2 mm layer thickness have maximum flexural strength (53.505 MPa).

5.4.5. Thermo Gravimetric Analysis (TGA)

TGA measures the amount and rate of change in the mass of a sample as a function of temperature or time in a con-trolled atmosphere. The measurements are used primarily to determine the thermal and or oxidative stabilities of materials as well as their compositional properties. It is useful for the study of polymeric materials, including thermoplastics, thermosets, elastomers, composites, films, fibres, coatings and paints. TGA provides information that can be used to select materials for certain applications.

A TGA consists of a sample pan that is supported by a precision balance. That pan resides in a furnace and is heated during the experiment. A sample purge gas controls the sample environment, it flows over the sample and exits through an exhaust. Mass of the test sample is monitored during the experiment.

A TGA thermal curve is displayed from left to right. The descending TGA thermal curve indicates a weight loss occurred. Temperature range from ambient 100 °C to 900 °C.

TGA results shows that FDM processed ABS sample un-dergo thermal degradation at temperature beginning at 389.57C with a mass loss of 89.19% and remaining portion is 7.98%and that of injection moulded sample undergo thermal degradation at a temperature begining at 390.48C with a mass loss of 88.17% and remaining portion is 8.759%. These two TGA graph shows similarities.but the TGA graph of the FDM processed sample shows closeness to the ideal curve of ABS than injection moulded sample.

CHAPTER 6

6. Research Trends in FDM of Thermoplastic Polymer - PLA

6.1. Introduction

PLA plastic or polylactic acid is a vegetable-based plastic material, which commonly uses cornstarch as a raw material. This material is a thermoplastic aliphatic polyester and it is the primary natural raw material used in 3D printing. PLA is a fully biodegradable thermoplastic polymer consisting of renewable raw materials. Among all 3D printing materials, PLA is part of the most popular materials used for additive manufacturing. Figure 6.1 shows a single PLA polymer monomer unit.

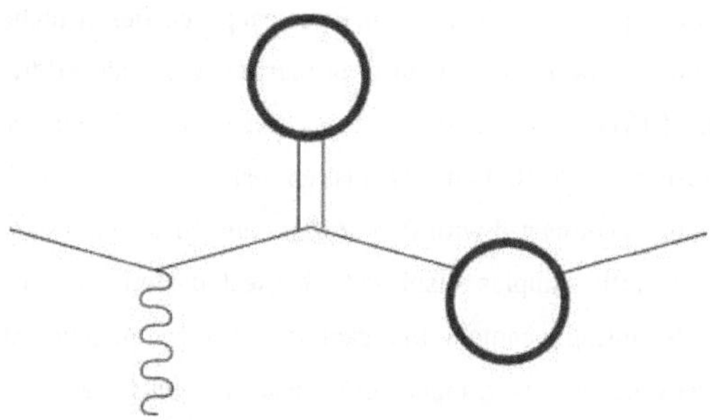

Fig. 6.1: Single PLA Polymer Monomer Unit [136]

PLA is used in 3D printing using the FDM (fused deposition modeling) technology, and along with ABS (Acrylonitrile Butadiene Styrene), this material is one of the standard materials for this technology.PLA plastic material is easy-to-use. Moreover, different versions of PLA filaments have been developed over the years. That is why you can now find some Aluminum PLA, PLA made with wood fibers or PLA with bronze particles. Possibilities offered by PLA actually seems to be endless.

In contrast, PLA will be less durable, more fragile, and more sensitive to heat than ABS. PLA has a glass transition temperature of 65 °C and a melting temperature of 178 °C. PLA is not the most heat resistant material, this is why it is particularly suitable for decorative objects with no mechanical constraints.

PLA material has numerous advantages, and that is precisely why this is the most used plastic material for desktop 3D printers. PLA filaments come in a lot of different shades, styles that can fit many different applications. This plastic material is especially efficient.

PLA is up to now the most used bio plastic for 3D printing by FDM, intended to be used in regenerative medicine, mostly as scaffolds for tissue engineering. PLA is thermoplastic aliphatic polyester prepared from fossil fuels or derived from renewable resources such as cornstarch or sugarcanes, rendering it accessible and inexpensive. PLA properties are strongly influenced by even small amounts of enantiomeric impurities.

6.2. 3D Printing of PLA Composites through Fused Deposition Modeling

PLA has relatively low glass transition temperature (50-60°C), low toughness and weak heat resistance, which limits its applications. Scaffolds made only of PLA do not mimic sufficiently the native bone structure and they do not ensure properly the cell colonization or mechanical properties. For some uses, PLA needs to be mixed with other polymers or fillers in order to create materials with improved thermal and mechanical properties or higher biocompatibility for biomedical purpose. Good improvement of properties was achieved by adding 15 wt% of nano-hydroxyapatite (HA) to PLA. The composite was extruded in filaments and then 3D printed at a nozzle temperature of 220 °C. Long-term creep test revealed a superior hardness of the 3D-printed composite as compared with PLA scaffold and consequently an increase in creep resistance. However, both samples displayed identical delamination destruction, due to limitations of the 3D-printing technique that cannot ensure completely sinterization between the layers. This causes the air to be trapped between layers, which lead to creation of voids. As expected, in vivo tests made on mice showed no inflammatory reaction even after 2 months and a slow biodegradation rate. Corcione et al. used filaments made of PLA and HA in different concentrations to obtain a molar tooth as shown in figure 6.2 this was successfully printed using FDM.

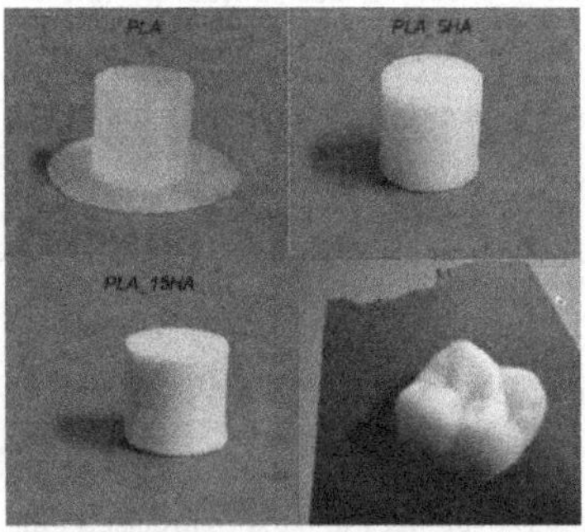

Fig. 6.2: PLA/HA Nano Composites by FDM 3D Printer [137]

Zhuang et al. used 3D printing to obtain plastic items with anisotropic heat and resistance distribution, which allows storing a simple message as color information in the printed objects. These were obtained from conductive graphene doped poly(lactic acid) (G-PLA) [138]. The authors used a method of programmed mixed printing to manufacture PLA composites with anisotropic properties. They stated that the method could be applied to other polymeric materials for a wide range of applications including biomedical ones.

For some medical applications, the rigidity and brittleness of the PLA are undesirable and the addition of elastomers is the easiest solution to overcome this drawback. TPU are among the most used polymers in 3D printing. They are also attractive for some biomedical applications due to their biocompatibility, high elongation at break and good abrasion resistance. As shown before, the use of different fillers impart to PLA exceptional mechanical strength, electrical conductivity, and enhanced thermal stability. Among them, composites with carbon fibers and graphene oxide (GO) proved to be also suitable for 3D-printing process. The addition of GO and TPU may have a cumulative effect of increased flexibility and mechanical strength. Chen et al. studied both the influence of the GO concentration and printing orientation on the mechanical properties of a TPU/PLA (7/3) blend [139]. Compression modulus tests have shown an increasing trend with the increase of GO content from 0.5 to 5 wt. % for both printing orientations, but the highest values were found for the specimens having the same printing orientation and height direction. The addition of only 0.5 wt. % GO determined an increase of the tensile modulus by 75% as compared with TPU/PLA sample, further addition of nanofiller determining a reduction of properties. This was explained by the percolation effect, which appeared below 2 wt. % GO content. All TPU/PLA/GO scaffolds supported fibroblast cells growth and proliferation, with the optimum effect at 0.5 wt. %.

6.3. Other Directions in 3D Printing of PLA based Materials

PLA may also fit the requirements for electronic devices and other fields by chemical modification or by the addition of different fillers and polymers. The presence of ionic liquids (IL) in a PLA 3D-printed structure provides unique features to PLA-based electronics; IL was recently added in the process of additive manufacturing of PLA filaments. The mixture was prepared by simply adding IL (5 and 10 wt. %) into a PLA chloroform solution, stirring for 12 h and then casting on a teflon plate. A significant enhancement of the PLA conductivity was noticed after the addition of trihexyl tetradecyl phosphonium decanoate, but further mechanical investigations are required to certify that this mixture is suitable for different applications.

Prashantha and Roger studied the mechanical and electrical properties of 3D-printed specimens made from commercially available PLA filaments filled with 10 wt. % graphene. The porosity distribution of the structure and the adhesion between layers were characterized through X-ray computed tomography. The results suggested that a shorter deposition time is favorable to obtain better interactions between the fused filaments and the maximum concentration for a suitable graphene dispersion is 10 wt. %. The increase of the electrical resistivity of the 3D-printed specimens, compared with the same composite before FDM processing, was explained by the alignment of the graphene nanoplatelets in the same direction with the deposited filaments. The reinforcing effect of graphene was highlighted by the increase of the storage modulus with more than 20% and tensile strength with 27%, with respect to PLA, as revealed by the DMA and static mechanical analysis [140].

PLA reinforced with 15 wt.% short carbon fibers (length about 60 mm) was manufactured by 3D printing based on fused filament fabrication and tested for mechanical and morphological properties. The PLA composite showed a higher increase in stiffness in the direction of printing. This behavior was explained by the morphological results, which revealed that the short carbon fibers were mostly aligned with the length of the 3D-printing filament, and remained aligned with the direction of printing within the PLA composite. Wood pulp fibers (WPF) are valuable reinforcements for many polymers but the application of FDM technology for 3D printing of biocomposites with WPF is a difficult process. The issues are related to the low thermal degradation temperature of the fibers, the small size of the nozzle used in FDM process and the poor dispersion of the fibers in the hydrophobic matrix, which causes fibers accumulation in the nozzle. A full enzymatic treatment was used to modify the surface of thermomechanical pulp (TMP) fibers; TMP fibers modified via laccase-assisted grafting of octyl gallate (OG) showed improved interfacial adhesion with PLA and a remarkable impact on the mechanical properties of PLA-TMP fibers composites. Moreover, filaments obtained from PLA reinforced with OG-treated fibers showed a good behavior during the 3D printing.

3D printing of a recycled PLA composite has proved to be a viable solution to the environmental issues, since the remanufactured 3D structure showed even better mechanical properties than the original one. Tian et al. have managed to recover a PLA/carbon fiber composite in a 100% rate for the carbon fiber and 73% for PLA matrix. They reused the material for the fabrication of new filaments, with a carbon fiber content of 10 wt.%, that were further processed by 3D printing [141]. No increase of the tensile strength was observed for the remanufactured composites as compared with the original composite, but other representative characteristics were improved, such as flexural strength, which increased with around 25%.

The aging process of the PLA matrix was impossible to be avoided due to repeated thermal cycles, but the mechanical performances were maintained by the addition of pure PLA in the 3D printing of the recycled composite.

6.4. Influence of Polymer Formulation on the Printing Quality and Mechanical Properties

Gianluca et al selected three different commercial polylactide (PLA) filaments for printing a challenging 3D geometry with overhang printing. Three industrial PLA-based filaments for FDM printing were purchased from different commercial sources. The three materials had different colors: white for the PLA Ultra by Roboze (Bari, Italy); black for the PLA Layer by Stampa3DSud (Salerno, Italy), and green for the PLA by Stampa3dSud (Salerno, Italy). The printing efficiency of the three filaments was tested by printing a complex geometry (i.e., Yoda face) without the use of a support material as shown in figure 6.3. These test trials are routinely used to assess the so-called overhang printing behavior. The FDM machine Zortrax M200 (Zortrax, Olsztyn, Poland) was used for all of the printing trials.

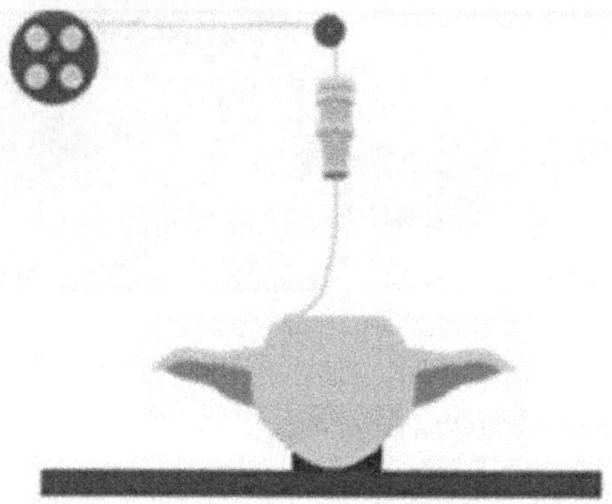

Fig. 6.3: Schematical Representation Showing the YODA sample While Printing [142]

The three PLA filaments were used to print the Yoda sample without the use of any support and by using always the same g-code. The Yoda shape has some critical printing issues with some selected parts, such as the ear and the chin, because of the absence of support materials. Such printing is usually referred to technically as overhang printing. These parts are critical, because the flow of the melted polymer with no control of sag properties can result in printing defects as shown in figure 6.4. These defects resemble those that are commonly reported as blobs and zits.

In our Yoda-printed samples, printing defects occurred only with the green PLA, whilst by using both the white and the black PLA, no visible defect was observed. The white Yoda shows no defect (a,c), the green Yoda showed printing defects in the critical area such as ears and the chin (b,d)

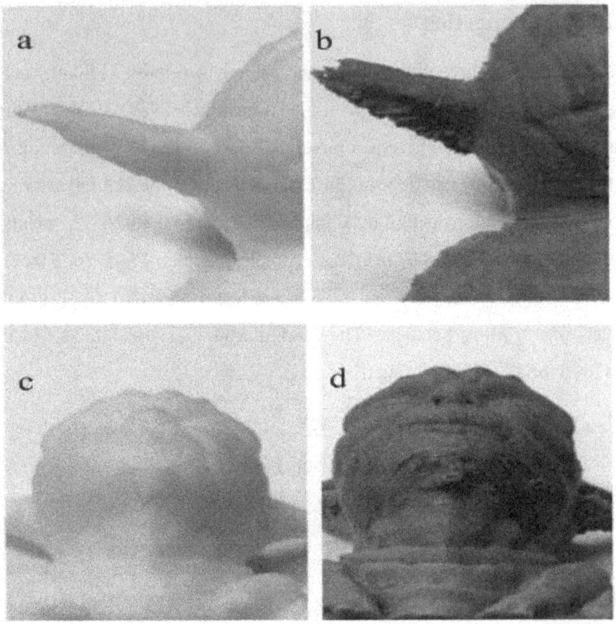

Fig. 6.4: Overhang Printing Defects Due to Uncontrolled Flow of Polymer [142]

6.4.1. Rheological Properties of PLA Filaments

An approximated estimate of the shear rate experienced by the polymer when passing through the nozzle leads to g of an order of 10^4 1/s. Capillary rheometer experiments were performed to access such a high shear rate regime. The results are shown in Figure 6.5. Since the black and the white PLA exhibited very similar rheological behavior, the data for the black sample are not shown for the sake of clarity. The three polymers exhibited comparable values of the viscosity for $\gamma > 10^3$ 1/s, sharing the same power-law behavior, h $\approx \gamma^{-0.78}$. On the other hand, noticeable differences emerged when lower shear rates were considered. In this regime, the data were collected via rotational rheometry. The white (and black) sample exhibited g-independent viscosity at $\gamma < 10^{-3}$ 1/s ($n_0 = 7 \times 10^4$Pa.s), whilst h rapidly decreased above this threshold.

Differently, the green PLA exhibited Newtonian behavior up to $\dot{\gamma} \sim 10^0$ 1/s, with a viscosity that was two orders of magnitude lower than that of the other two samples. This difference is likely correlated to the different process ability of the materials in the course of FDM.

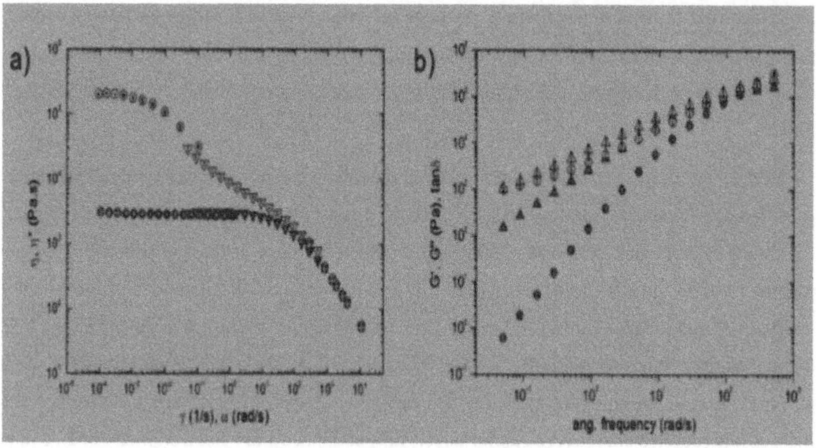

Fig. 6.5: (a) Shear (Circles) and Complex (Triangles) Viscosity for White (Open Symbols) and Green (Full Symbols) Polylactide (PLA). (b) Elastic (Circles) and Viscous (Triangles) Modulus for White (Open Symbols) and Green (Full Symbols) PLA [142]

6.5. Surface Quality in FDM Process

In additive manufacturing, in general, pre-processing and post-processing activities should be carried out. However, the quality of the parts is not adequate when compared to other mature manufacturing processes, such as machining. One of the main problems for obtaining good surface quality in additive manufacturing is the staircase effect. Usually manual post-processing operations are needed for obtaining adequate surface roughness because complex geometries compromise the advantages of additive manufacturing. Various studies have been specifically carried out on FDM process parameters, discussing their effect on outputs, such as mechanical properties, and surface topography and quality. For instance, Altan et al studied the effect of process parameters on surface roughness and the tensile strength on polylactic acid (PLA) samples. The samples were fabricated as per the ASTM standards and a Taguchi L16 experimental design, using three parameters: layer thickness, deposition head velocity, and nozzle temperature. The authors concluded that the layer thickness and deposition head velocity are dominant factors on surface roughness [143].

From previous scientific records, it has been identified that the performance of the FDM process extensively depends upon its process parameters and their levels. However, the number of publications dealing with the identification of critical factors and the optimization of the manufacturing process depending on adequate selection of factors and levels is still limited, particularly when it comes to surface roughness. The optimization of the process parameters used for printing is an adequate strategy for improving part performance in terms of surface finish.

Mercedes et al in their experimental worked on enhancement of surface qualities based on selection of critical parameters. The 3D printer used was a WITBOX printer as shown in figure 6.6 by BQ manufacturer (Madrid, Spain) using FDM technology. It is equipped with 0.4 mm diameter nozzle and glass cold base, A4-size (297 _ 210 mm). The recommended printing speed is 50 mm/s and the maximum is 80 mm/s. The own-design extruder has a blower for cooling printed objects. Polylactic acid (PLA) (Smart Materials 3D, Alcalá la Real, Spain) has been used as base material. The company that provided the material was Smart Materials 3D. It does not incorporate recycled or recovered material. It is fully stabilized and it has a diameter of 1.75 mm with a variability of 0.03 mm in diameter. Besides, no warping is expected.

Fig. 6.6: WITBOX Printer and PLA Filament [144]

Surface roughness was measured using a handheld Mitutoyo Surftest SJ-210 (Mitutoyo, Kawasaki, Japan) profilometer as shown in figure 6.7 with sampling length of 2.5 mm and measuring speed of 0.5 mm/s with this technology, cylindrical samples such as those shown in Figure 6.8 were measured.

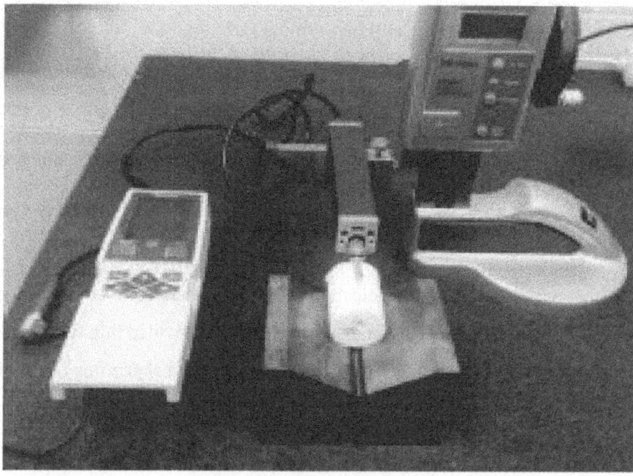

Fig. 6.7: Surface Roughness Measuring Setup [144]

Fig. 6.8: Printed Samples for Experiment [144]

6.5.1. *Experimental Plan*

In additive manufacturing, there are several factors that could influence surface roughness, such as:

- Material extrusion: temperature, viscosity, density, type of material, and mechanical properties.
- Chamber: temperature, pressure, vibrations, position of the platform, position of the extruder, system coordinates, and heat evacuation.
- Extruder: speed, angle of inclination, diameter of extrusion, vibration, and acceleration.
- Deposition characteristics: building direction, wall thickness, layer height, orientation, external geometry.

Due to the high number of factors, a selection of factors to carry out a more economical and practical study was made. The experimental investigation was divided into two stages. The first stage was designed as a screening stage to identify the most critical printing factors for surface roughness. The second stage was performed in order to increase the knowledge of the printing factors based on the results of the first stage.

All printed samples had dimensions of 30 mm in diameter and 40 mm in height. The first analysis was done to study the influence of layer height, wall thickness, printing speed, and temperature (material). These factors were varied using two levels: minimum and maximum. So, eight tests were performed by means of a fractional factorial design of four factors with two levels. Fractional factorial designs allow carrying out experimental studies with limited number of experiments and thus, reducing cost and time.

For layer height, values of 0.15 and 0.25 mm were chosen. The first one is the minimum recommended by the predefined options of the Cura software. The second one is a higher value, which was selected expecting an increase in surface roughness as it was identified in the literature. For printing speed, a value of 40 mm/s was selected; a speed lower than that recommended by the printer manufacturer, and a value of 80 mm/s, the maximum recommended. For temperature a maximum value lower than the one recommended by the PLA filament manufacturer (240^{0}C) was selected, i.e., 225^{0}C and, as minimum value, 195^{0}C was selected that lies slightly below the minimum recommended (200^{0}C). Finally, values of 1 and 3 mm were selected for wall thickness, considering that wall thickness should be higher than two times the size of the nozzle extruder.

The second analysis was done to specifically study the influence of wall thickness and its relation to the printing path: zig-zag, concentric, and grid. These tests were performed using the parameters used in the first stage to obtain one of the best surface roughness, so the lowest layer height was chosen (0.15 mm), but it was decided to also have a reduced printing time (estimated printing time of 47 min), so the printing speed of 80 mm/s was selected. Moreover, temperature of 225⁰C was chosen. So, two factors were analyzed in this stage, using three levels for printing strategy and five for wall thickness. For printing path, concentric, zig-zag, and grid were selected.

6.5.2. Surface Roughness Evaluation

Surface roughness was evaluated in terms of the arithmetic average of the roughness profile (Ra). Six surface roughness measurements were taken in each sample. The samples were divided into two sections: bottom (printing start) and top (printing end) sections. In addition, three generatrices were drawn on the surface. No measurement was done in the section where the zipper effect generated by the layer change can be seen. The measurements of Ra1, Ra2, and Ra3 were taken at the top in the generatrices in a clockwise direction. The measurements of Ra4, Ra5, and Ra6 were taken at the bottom in the generatrices in a counter clockwise direction shown in figure 6.9. Therefore, 48 and 90 measurements were obtained for experiment 1 and 2, respectively. Finally, with the six surface roughness measurements, the average roughness was calculated for each sample.

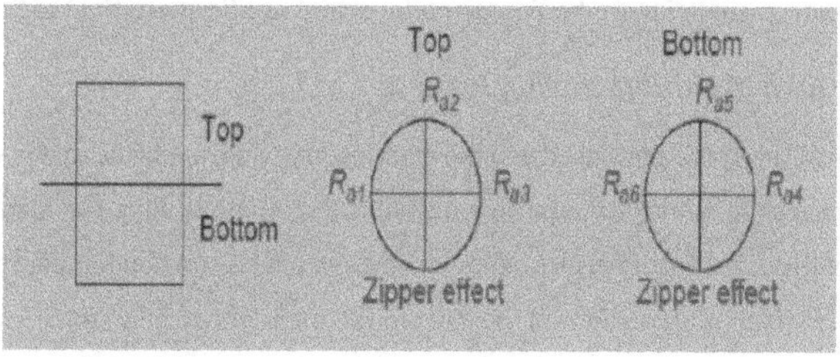

Fig. 6.9: Details of Measurement Procedure [144]

6.5.3. Experimental Findings

The main conclusions of the investigation include the following:

The quality of the manufactured parts depends greatly on the selection of the printing parameters. In particular, previous results that indicate that the layer height is a critical factor were validated using Analysis of Variance. But, in addition, it was found that wall thickness has an important influence on the results, which was given less attention by researchers.

- Some parameters seem to have no clear influence on the results, specifically, printing path, printing speed, and temperature. However, it should be noted that only three printing strategies were analyzed in the present study: grid, concentric, and zig-zag.
- By using Spearman's r and Kendall'scorrelation coefficients, the influence of layer height and wall thickness on the results was verified, especially, for experiment 2, obtaining correlation coefficients very close to +1 with p-values lower than 0.05.
- The effect of the layer height and wall thickness on surface roughness is to worsen the quality of the surface when one of these parameters is increased or when both are increased.
- As criteria for improving surface quality in FDM manufacturing processes, it is recommended to use reduced values of layer height, diminishing the importance of the staircase effect and also wall thickness that is generally selected based on the size of the nozzle extruder.

6.6. Tensile Properties of PLA Specimen

Cristina et al investigated the tensile properties of PLA printed specimens obtained through FDM technology and observed that the spatial orientation has less influence on the Young modulus and higher influence on the tensile strength. Furthermore, increasing the number of layers leads to decreasing of both the Young modulus and tensile strength [145].

6.6.1. Material and Manufacturing Process

Polylactic acid (PLA) is a bio-based and bio compostable thermoplastic that is widely used in different industries for due to its superior mechanical strength. Unlike the high strength and high stiffness of this material, its brittle behavior and low heat distortion temperature have been pointed out to be its limitations in real life. Due to the mentioned characteristics of PLA, it has been widely considered as a model material in FDM printing. WN400 3D platform printer was used for fabrication of the test specimens. The printer was equipped with an HFE300 extruder for printing parts with filaments of 2.85 mm diameter. 3D printing software was used to set the printing parameters such as raster angle, head speed, temperature and so on. To ensure the quality of the printed part, the temperature of the nozzle and the built platform was controlled at around 60°C and 220°C, respectively. Was defined an infilldensity of 100% in the printing software with raster angles of ±45 degrees for the infill and different layer thicknesses were considered to print layers.

Following the 3D printing process, dog-bone (DB) specimens were obtained. The geometrical parameters of the DB specimens followed the ISO 527-1 (2012) standard recommendations as shown in figure 6.10.

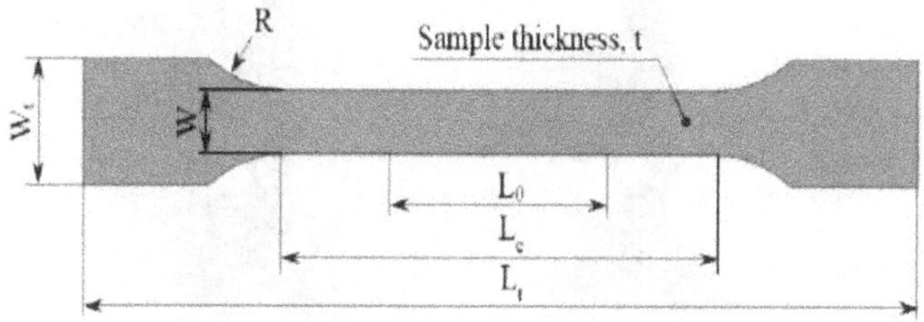

Fig. 6.10: Geometrical Parameters of DB Specimen [145]

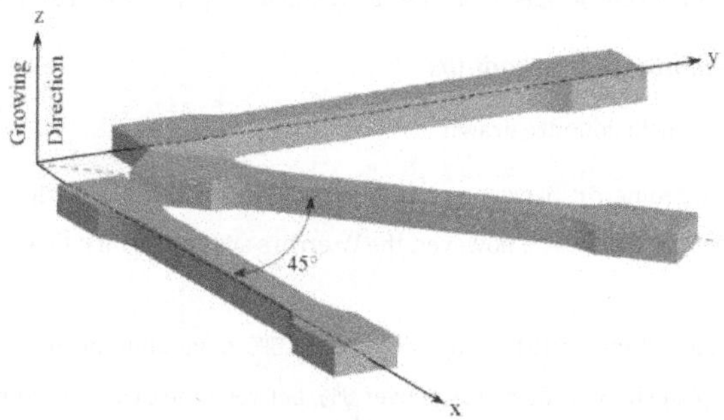

Fig. 6.11: Printing Orientation of 3D Specimen [145]

In order to obtain the main geometrical and tensile properties of the FDM printed specimens, different printing orientations (PO) were used. Figure 6.11 shows a schematic view of the PO (0, 45 and 90°). In the same figure, the growing direction is presented; therefore, all specimens were manufactured in a horizontal plane.

6.6.2. Tensile Test

Tensile tests were performed on a 5 kN Zwick Roell 005 electromechanical testing machine, according to ISO 527- 1 (2012) standard. All tests run up to the failure point with a loading speed of 2 mm/min, at room temperature. A clip-on extensometer, with calibrated distance 30 mm was used for measuring strains, instead of the crosshead displacement of testing machine. For each type of specimen were tested a number of five specimens. Figure 6.12 shows the DB specimen in the tensile testing machine grips together with the positioning of the extensometer, both before and after the experimental test.

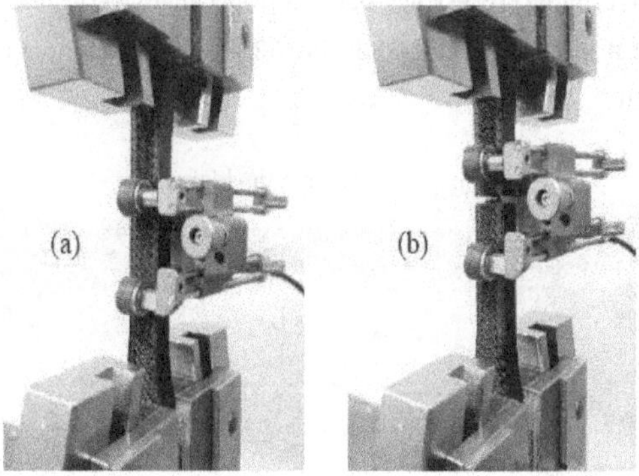

Fig. 6.12: Tensile Specimen Before (a) and After (b) Tensile Test [146]

6.6.3. *Experimental Work Findings*

The following conclusions are drawn:

- The main geometric parameters (thickness-t and width-W) of the specimens have relative errors below 4%; however, the W errors are approximately double that of the t ones.
- The Young's Modulus (E) changes by only 1.8% depending on the PO, while tensile strength (σ_m) shows differences of over 8%, between the extreme values. Regardless of the PO, E and σ_m have errors below 1.5%.
- Both E and σ_m decrease significantly and polynomial with increasing sample sizes (30% for E and over 7% for σ_m). Therefore, due to the presence of the inherent defects, a strong size effect is identified, especially for E.
- The determined values of tensile strength for PLA obtained through AM are in good agreement with those from injection molded, except of 45^0 orientation which are a little bit lower.

6.7. Compression Test of PLA Specimen

Salvatore et al in their experimental work investigated the compression properties of pla specimens. The physical properties of these element were difficult to be determined and they depended on the deposition temperature, deposition rate and layer thickness used in the printing process. For compression tests black PolyLactic Acid-PLA (produced by Shenzhen Eryone Technology Co., Ltd., Shenzhen, China) and they were produced via the 3D Fused Deposition Modelling (FDM) technique.

A professional 3D printer was employed for the production of specimens for compression tests, it was the Funmat HT (this 3D printer is based on the Fused Filament Fabrication (FFF) technology which is often also called FDM, even if this last one is sometimes considered as a peculiarity of the Stratasys devices). The compression specimens were parallelepipeds with a square cross section and an opportune length in order to avoid buckling phenomena (see the Standard Reference ASTM D695 for compressive tests).

6.7.1. Specimens for Compression Test

The specimens for compression tests had the shape and the main geometrical dimensions as indicated in Figure 6.13.

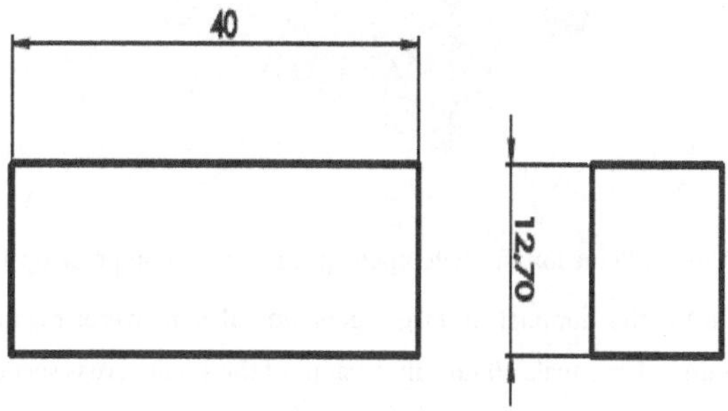

Fig. 6.13: Geometrical Data (in mm) of the Parallelepiped-shaped Specimen for the Compression Tests [146]

The 3D CAD draw was converted in the STL format. The parallelepiped-shape specimen had 6 faces and for each face two triangles were sufficient to describe all the internal and external surfaces. Therefore, 12 triangles were enough for the triangulation of the entire parallelepiped-shape specimen. The slicing operation allowed the conversion of the STL file in a G code format in order to give the main instructions to the Funmat HT printer for the production process. As clearly indicated in figure 6.13. The printing plane as the xy one and the layers were summed along the z direction; the same _45_ criss-cross sequence with rectilinear infill already employed for specimens for tensile tests was here used. The 12 specimens were printed in two different periods by giving two families composed by 6 specimens per each one. The main printing data were the same already seen for specimens for tensile tests, the only difference was for the perimeter beads. In the present case, no perimeter beads were used because of the simplicity of the geometry and also to avoid unwanted reinforcement phenomena in the compression direction.

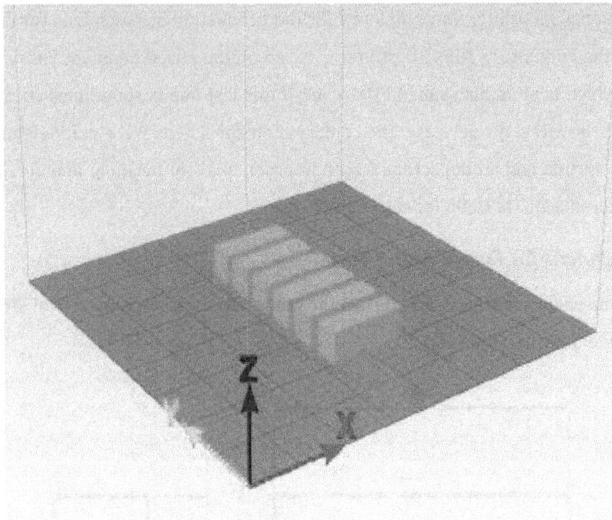

Fig. 6.14: Printing Plane for Parallelepiped Specimens for Compression Tests [146]

From Figure 6.14, the nominal or target geometrical data were: overall length of the specimen in the x direction equals 40mm, dimensions of the square cross section were 12.7mm in both directions y and z. The target value for the weight was 8.06 g and it was obtained by considering the volume of the specimen (as given by the CAD model) multiplied for the infill density (equals 1) and for the mass density of the PLA equals 1.25 g/cm3. The real geometrical data for each produced specimen were measured by means of a digital caliper and the weight of each specimen was determined using a digital weight scale.

6.7.2. Results for Compression Test

A standardized test method for FDM 3D printed polymers has not been developed yet; therefore, the ASTM D695 for classical polymers has been employed. A constant speed for the downward movement of the upper crossbar is requested, this speed is here equalto -1.8mm/min. In the case of the compression test, w is the width in millimeters of the square cross section of each tested specimen in the Y direction, t is the width in millimeters of the square cross section of each tested specimen in the Z direction and L0 is the length of each specimen in the X direction in millimeters (all these data are the actual values of each specimen). The slenderness ratio is the ratio between the free length of inflection or effective length of the specimen and the least radius of gyration of its cross section. The standard reference suggests values between 11:1 and 16:1 in order to avoid buckling phenomena.

The specimens here printed have a slenderness ratio value roughly equal to 11:1. In fact, the effective length is equal to 40mm (the presence of the two plane plates can be considered as simply supported boundary conditions) and the least radius of gyration of a square cross section is equal to 0.289w. The figure 6.15 given below shows the stress- strain curves of the performed compression test.In each image, only the linear elastic Young modulus Elin in MPa is given because the tests were stopped when the machine MTS QTest arrived to the maximum applicable load P equals 10kN (which is the value of the employed load cell). For this applied load, only a linear elastic region is clearly showed and no proportional limit stresses and ultimate tensile strengths can be defined in an appropriate way.

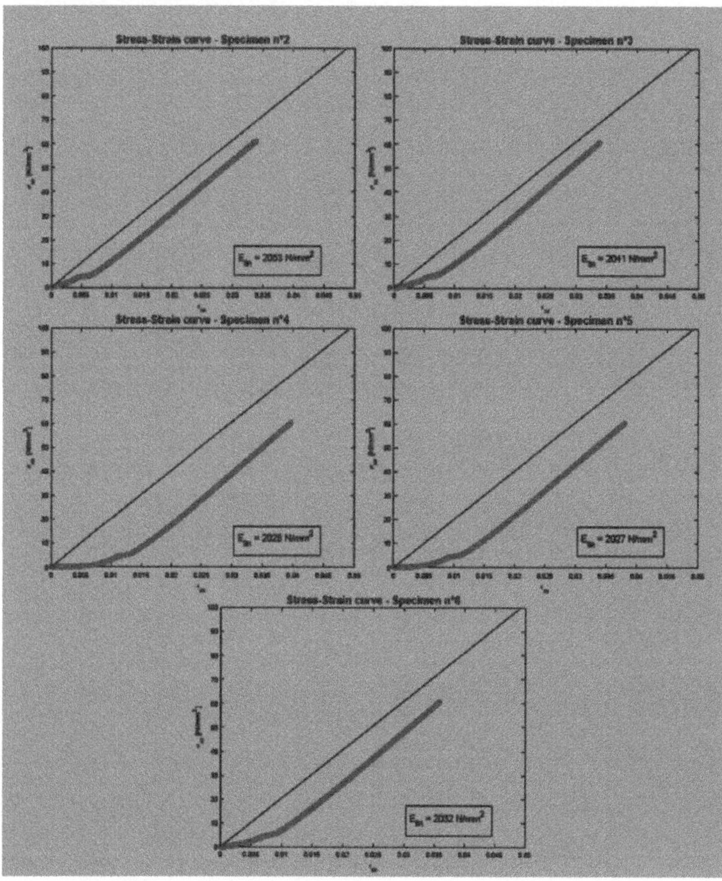

Fig. 6.15: Stress-strain Compression Test Curves for the Test Specimens [146]

The black line represents the linear elastic region and its slope is the linear compression Young modulus obtained from a linear regression. In each image of figures shown above, a toe point is clearly shown, which is a point with a horizontal tangent between two different linear elastic regions. This toe point is due to a specimen misalignment, which is generated by a slipping in the grips, with a consequent realignment with respect to the load application direction. This phenomenon is a mechanical test characteristic and not a material characteristic.

CHAPTER 7

7. Research Trends in Ultem Polymer

7.1. Introduction

Polyetherimide (PEI) with the trade name Ultem 9085 is an amorphous and transparent polymer. This material is desirable due to its mechanical properties, relatively low density compared to traditional materials, and flame, smoke, and toxicity properties that allow its use in aircraft cabins. The material PEI is used on FDM machines for the manufacturing of end products. It shows high glass transition temperature, excellent flame retardancy, low smoke generation, and good mechanical properties. This set of properties makes this material a promising candidate in different application fields, especially those which require low mass and high strength such as aerospace, marine and automotive sectors. As Ultem 9085 is a relative recent material, no extensive literature is available. There are however, two main aspects that are more deeply investigated in the literature due to their large impact on the mechanical performance of printed parts. Firstly, the quality of final components is greatly influenced by the used-process parameters. FDM involves a quite complex mechanism based on the interaction of different parameters such air gap, raster width, raster angle, contour number, and contour width. Their optimization and the study of how these parameters can individually or collectively affect the properties of printed objects is fundamental to obtaining high performing components. The promising performance of Ultem 9085 and its potential application in the transport sector, with particular reference to the automotive sector, generates interest in testing this material under severe environmental conditions.

7.2. Materials

Ultem 9085, a high-performance thermoplastic polymer specifically designed for FDM technology; it is supplied by Stratasys in the form of natural-colour filament with a diameter of 1.75 mm. This material was used for the preparation of specimens by using a Fortus 450mc 3D-printer supplied by Stratasys. The fabrication of samples consists of three main steps. Firstly, the test coupons were modelled using the commercial software Solidworks, and then a machine code in STL (stereolithography) format was generated. Secondly, the STL file was exported into the software package Insight 12.1, which was supplied with the AM machine. It was used to set the build parameters and control all the printing stages.

All the samples were fabricated by using the same building parameters: raster width of 0.508 mm, contour width of 0.508 mm, air gap of 0 mm, and contour number of 3. Moreover, filling of layers (a theoretical 100% infill was set) was performed alternating raster angles of ±45 ◦ with respect to the x-axis. Finally, the STL file was sent to FDM machine, which began to fabricate the specimens by extruding Ultem 9085 filament and depositing it layer by layer. The scanning strategy involved the deposition of the contour of a single layer followed by its filling according to the prefixed raster angle. The specimens for tensile and bending tests were manufactured according to XY, XZ, and ZX building orientations as shown in figure 7.1, where the first letter specifies the direction of the main axis of the specimen with respect to the build platform, and the second letter, together to the first one, identifies the plane on which the largest sample surface lies.

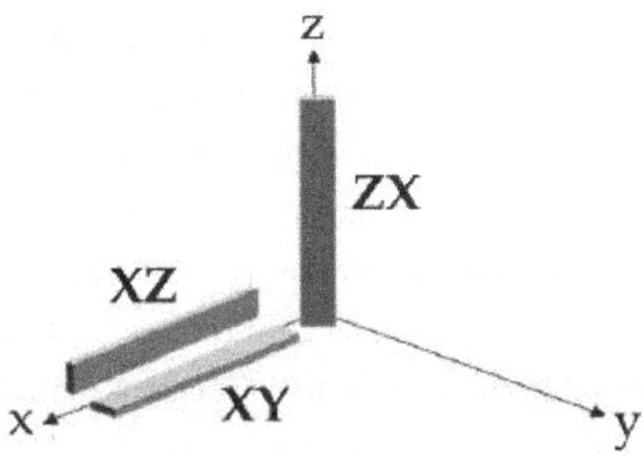

Fig. 7.1: FDM Samples Built Direction [147]

7.3. Density

The measurement of density was performed on the Ultem 9085 in the form of both filament and printed components. The results are reported in Table 7.1. The density of the filament was found to be equal to 1.2864 ± 0.0005 g/cm3; this value was taken as reference to calculate the relative density of printed samples as well as the porosity values.

Table 7.1: Average Value of Densities and Porosity of Printed Parts for the different Build Orientation [147]

Building orientation	**Density (g/cm³)**	**Relative density (% of Theoretical)**	**Porosity (%)**
XY	1.2389 (0.0004)	96.40 (0.03)	3.60 (0.03)
XZ	1.2552 (0.0003)	97.30 (0.02)	2.69 (0.02)
ZX	1.2637 (0.0002)	97.96 (0.02)	2.04 (0.02)

From Table 7.1 it is evident that samples showed increasing relative density in the printing directions XY < XZ < ZX; as expected, the corresponding values of calculated porosity showed a reverse trend. This trend of density variation when different printing directions are considered agrees with that observed by Byberg et al [148]. The variation of density observed for the different building orientations is mainly influenced by the ratio between the extent of the contour and the infill zones, and the scanning strategy adopted to fill each layer. The extruder head firstly deposits the contour of the object outlining its perimeter; then, it completes the layer by depositing the filament within the layer contour with an orientation of +45 ∘ /−45 ∘. The misalignment of the infill with respect to the contour causes the formation of voids in fact, when the +45 ∘ /−45 ∘ filament meets the contour and goes back towards the opposite side, an empty space is left at the corner of infill filament.

7.4. Tensile Test

Tensile tests were performed on dog-bone specimens with building orientation XY, XZ, and ZX, respectively. The statistical experiment was carried out using the one-way analysis of variance (ANOVA) in order to evaluate the main effects of building orientations (the independent variables) on each tensile property under investigation (yield strength, ultimate tensile strength, elastic modulus, and elongation at break, which are the response variables). Specifically, the p-value which indicates the level of significance of the different factors within a statistical test is reported; it represents the probability of a factor affecting the mechanical properties. The significant factors are tested with p-value lower than 0.05. The statistical analysis table 7.2 revealed significant differences among building orientations for all the tensile properties under investigation; this implies that at least one of them differs from the others. The Tukey's HSD post hoc test (α = 0.05) was performed in order to identify which pairs of building directions are significantly different from each other.

Table 7.2: ANOVA Test for Tensile Properties [147]

Mechanical property	Source	Sum of Squares	F	F Crit	p-value
Yield strength	BG	131.961	750.954	3.885	<0.001
	WG	10.722			
Ultimate tensile strength	BG	131.961	120.109	3.885	<0.001
	WG	10.722			
Elastic modulus	BG	74.080	4.206	3.885	0.040
	WG	105.680			
Elongation at break	BG	109.256	95.608	3.885	<0.001
	WG	6.856			
BG=between groups, WG=within groups, SS=sum of squares, F=F ratio, F crit=critical F ratio					

Table 7.3 shows the means and the standard deviations for the tensile properties such as yield and ultimate tensile strengths, elastic modulus, and elongation at break; moreover, it compares these outcomes with those present in the technical datasheet provided by Stratasys (in this case, specimens were built by using Fortus 3D printer setting default process parameters defined by Stratasys).

Table 7.3: Results of Tensile Properties of Ultem9085 Samples Built According the Three Building Directions [147]

Mechanical property	XY	XZ		ZX	
	Experimental	Experimental	Datasheet	Experimental	Datasheet
Yield strength [MPa]	47.0 (0.6)[b]	54.8 (0.3)[c]	47	32.0 (1.5)[a]	33
UTS [MPa]	65.9 (0.7)[b]	73.0 (1.3)[b]	69	39.6 (6.0)[a]	42
Tensile modulus [MPa]	2220(54)[a]	2300 (54)[a]	2150	2128 (144)[a]	2270
Elongation at break [%]	6.6(0.6)[b]	8.0 (0.2)[b]	5.8	1.7 (0.8)[a]	2.2

It is evident that XZ orientation was the most performant one showing the highest values of yield and ultimate tensile strengths (UTS), elastic modulus, and elongation at break. However, pairwise multiple comparisons with Tukey's HSD test revealed that UTS, tensile modulus, and elongation at break of XY samples were not significantly different with respect to those determined for XZ specimens. On the other hand, ZX samples reported statistically lower tensile properties compared to the both XY and XZ building directions. Despite the value of tensile modulus for ZX samples not being significantly different from those obtained for the other two orientations, tensile strength and elongation at break were much lower with respect to those observed for XY and XZ directions. The observed trend for tensile properties is in good agreement to that reported in the literature [149,150].

Samples after tensile tests are shown in Figure 7.2 (a–c). These images provide evidence of the different filament deposition patterns adopted for the building of samples with XY, XZ, and ZX orientations, respectively. Figure 7.2 (d–f) show the fracture surfaces of the three kinds of samples. From a visual analysis it seems that ZX cross section shows lower porosity degree than XY and XZ samples. The presence of few voids can in fact be observed mainly in the region close to the interface between infill and contour (red arrows). Similarly, in XZ samples the porosities seem mainly located at the border area where infill and contour filament meet; however, the amount of observed voids seems higher than for ZX samples. Moreover, the visual inspection of XY surface fracture evidences the presence of some voids both at the interface infill/contour and in the infill area. These qualitative observations agree to the density values reported in Table 7.1.

Both XY and XZ specimens, which granted the best tensile properties, showed a brittle fracture. The slightly higher UTS of XZ samples with respect to the XY ones was probably due to the different contour/infill areas ratio. The observed fracture surfaces show a rectangular shape where most of the area is constituted by the infill, and only a minority part represents the contour. In both XY and XZ samples, the filament which forms the contours is deposited parallel to sample axes and, as a consequence, in the same direction of load application.

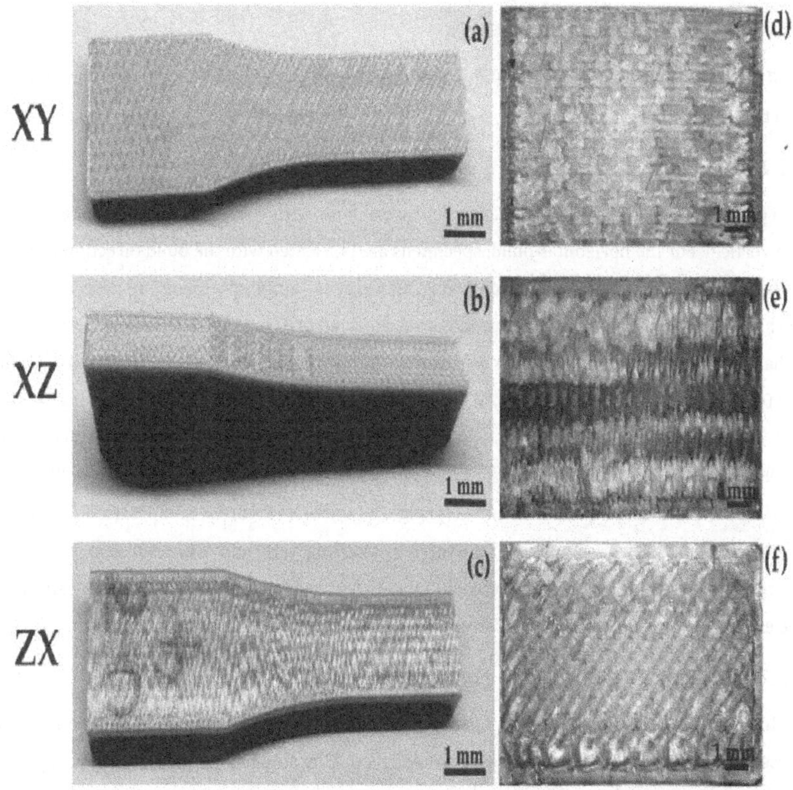

Fig. 7.2: Macroscopic Views of Samples (a–c) and Fracture Surfaces (d–f). [147]

This allows filaments to strongly oppose to the load application. On the other hand, the filaments which constitute the infill are placed at 45 ° with respect to the direction of load application.

In samples with XZ orientation, the contour/infill areas ratio is higher with respect to XY samples because of the contour is present on the longest side of the rectangular cross section (while in the XY samples, the contour represents the shortest side). This may explain both the higher strength and elongation of XZ specimens with respect to XY ones. In addition, the improved mechanical behavior of XZ samples with respect to XY ones can be supported by the higher degree of porosity observed in XY samples with respect to XZ ones; the porosity can in fact negatively affect the mechanical properties of the material.

7.5. Effect of Build Parameters

Krishna et al in their experimental work investigated the effects of various parameters of ULTEM 9085 [151].

7.5.1. Specimen Details

For this study, the Ultem 9085 specimens were built with two build directions, horizontal and vertical. For the horizontal-build, specimens are fabricated with the build direction parallel to the compression load to be applied later. For the vertical-build, specimens are fabricated with the build direction perpendicular to the compression load to be applied. Raster angle varied with either (45°, -45°) and (0°, 90°). Air gap varied with -0.00635 mm (-0.00025"), -0.0127 mm (- 0.0005"), or -0.01905 mm (-0.00075") for solid specimens but was held at a constant 2.54 mm (0.1") for the sparse specimens. Raster width and contour width were both held at a constant 0.508 mm (0.02") for all specimens. Specimens were fabricated with a 38.1 mm x 38.1 mm (1.5" x 1.5") cross-section and 25.4 mm (1") height as shown in Figure 2. Compression tests were performed with a 1.27 mm/min (0.05 in/min) loading rate and were stopped when the specimen had over 10% strain. Effective modulus was calculated as stress divided by strain in the elastic region and effective yield strength was calculated as strength at 0.2% offset.

7.5.2. Design of Experiments - Factors and Levels

In design of experiments, independent variables, also known as factors, and variations for these variables, also known as levels, are established. This study splits the specimens into two experiments, solid and sparse specimens.

For the solid specimen case, three factors including build direction, raster angle, and air gap were used (Table 1). Build direction has two levels, horizontal and vertical. Raster angle has two levels, (0°, 90°) and (45°, -45°). Air gap has three levels, -0.00635 mm (-0.00025"), -0.0127 mm (-0.0005") and -0.01905 mm (-0.00075").

While increasing the magnitude of the negative air gap in the solid specimens will generally increase the strength, the negative air gap will greatly decrease the surface quality of the specimens when the air gap reaches a certain limit (- 0.01905 mm for the Fortus 400mc). A full-factorial experiment was performed using the 12 combinations with replications of 5 samples for each combination.

Table 7.4: Factors and Levels for Solid Build Style [151]

Factors	**Levels (5 Replications)**		
	-1	0	1
Build direction	Horizontal	-	Vertical
Raster angle (degrees/degrees)	0/90	-	45/-45
Air gap (mm)	-0.00635	-0.0127	-0.01905

In the sparse case, two factors including build direction and raster angle were used Table 7.5. Unlike the solid case, the air gap was fixed at 2.54 mm (0.1"). Build direction has two levels, horizontal and vertical. Raster angle has two levels, (0°, 90°) and (45°, -45°). A full-factorial experiment was performed using the 4 combinations with replications of 5 samples for each combination.

Table 7.5: Factors and Levels for Sparse Build Style [151]

Factors	**Levels (5 Replications)**		
	-1	0	1
Build direction	Horizontal	-	Vertical
Raster angle (degrees/degrees)	0/90	-	45/-45

7.5.3. Effect of Build Style and Temperature

Five specimens were tested for each combination of solid/sparse, build direction, and temperature. Build styles included horizontal-build and vertical-build while temperatures included 24°C (75°F), 82°C (180°F), and 121°C (250°F). The obtained compressive modulus, yield strength, specific modulus, and specific yield strength of the specimens were averaged and plotted along with standard deviation.

From compression test results shown in Figure 7.3, at all three temperatures [24°C (75°F), 82°C (180°F), 121°C (250°F)], the yield strength, compression modulus, strength/mass ratio, and modulus/mass ratio of the horizontal-build specimens (S-H and SP-H) are higher in comparison to the corresponding vertical-build specimens (S-V and SP-V). The specimens with S-H build style (solid specimen with horizontal-build) exhibits the highest strength and modulus at each respective temperature. These results clearly indicate that both solid/sparse build and build direction affect the compression properties of FDM specimens.

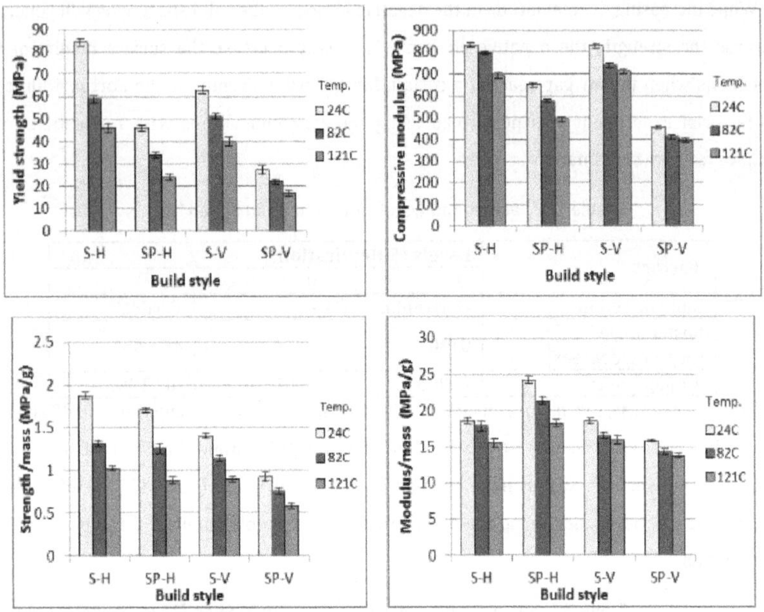

Fig. 7.3: Comparison of Compression Properties for Solid and Sparse Specimens [151]

The results show the yield strength of the solid specimen decreases with increase in temperature as illustrated in figure 7.4. In the case of horizontal-build specimens, the yield strength decreases more than the vertical-build specimens for each respective temperature. This indicates that the vertical build specimens display a relatively higher resistance to temperature in comparison to the horizontal-build specimens, whereas horizontal-build specimens display relatively higher strength.

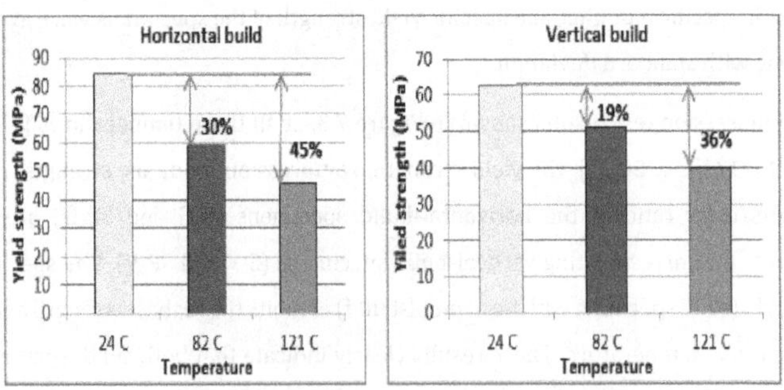

Fig. 7.4: Yield Strength of Horizontal Built and Vertical Built Solid Specimens [151]

7.5.4. Effects of Build Direction, Raster Angle and Air Gap

Experiment was designed to conduct compression tests on both solid specimens with varying build parameters including build direction, raster angle, and air gap. A full-factorial design of experiment was used for solid specimens and the effects of the parameters were studied. Compression tests were conducted on the FDM fabricated solid specimens. A total of 60 specimens were tested for 12 different combinations of solid build parameters and 5 replications for each set of parameters. The study investigated compression properties including modulus, yield strength, specific modulus, and specific yield strength. The full-factorial experiment was used for the determination of the main effects and the interactions among the parameters.

Main effect is defined as the effect of an independent variable on a response variable averaging across different levels of other independent variables. Independent variables include build direction, raster angle, and air gap for the solid specimen case. If the effect of one independent variable on the response variable is dependent on the value of another independent variable, those two variables are said to exhibit interaction. Yield strength was the only response variable considered during the experiment. All results were analyzed with JMP 11. Results from the statistical software are given in Table 7.6. The p-value defines the level of significance within a statistical test and represents the probability of a factor affecting the outcome (response variable).

Table 7.6: Effects Table from JMP 11 for Solid Specimen [151]

Factors	**Sum of squares**	**F ratio**	**p-value**
Build direction	4775.5037	8392.691	<.0001
Raster angle	343.6323	603.9153	<.0001
Air gap	6.9546	3.4946	0.0816
Build raster	249.6281	438.708	<.0001
Build*Air gap	1.5674	2.7547	0.1019
Raster* Air gap	0.7047	1.2385	0.2699

For this statistical experiment, if the p-value is less than 0.05 (for a significance level of 95%), the corresponding factor has a significant effect on the response variable. All p-values less than 0.05 are highlighted in red to indicate main effects and/or interaction. The effects table indicates an interaction between build direction and raster angle, which means that the effect of raster angle on the response variable is dependent on the build direction used for the specimen. The interaction of these parameters is described below in detail.

7.5.5. *Effect of Build Direction and Raster Angle*

The two different raster angles evaluated in this experiment included (0°, 90°) and (45°, -45°) for both horizontal-build and vertical-build solid specimens. The results are shown in Figures 7.5 and 7.6.

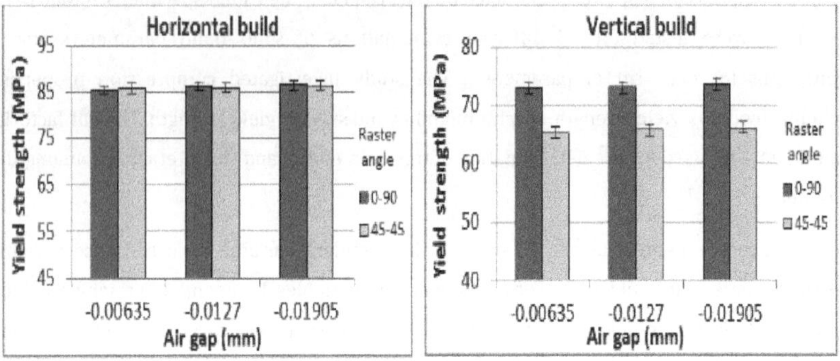

Fig. 7.5: Interaction of Build Direction and Raster Angle on Yield Strength of Solid Specimens [151]

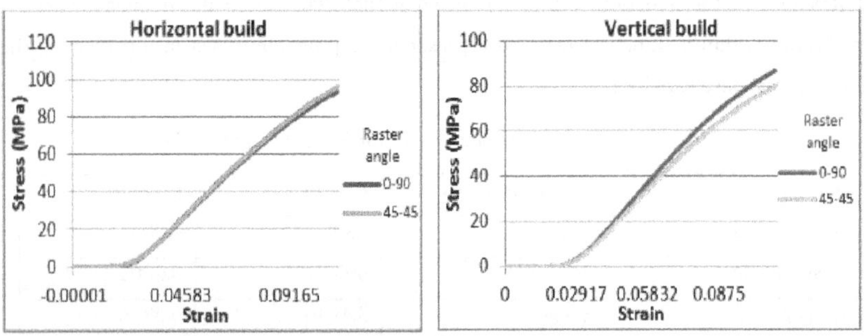

Fig. 7.6: Sample Stress-strain Curves for Horizontal and Vertical Build Specimens [151]

Horizontal Build

According to the graphs in Figures 7.6 and 7.7, the horizontal-build specimens with (0°, 90°) and (45°, -45°) raster angles have very small differences in the yield strength, and the trend remains the same for different values of air gap. This occurs because in the case of horizontal-build specimens, the applied load is perpendicular to the layer in which the raster's are present, and the internal structure for resisting the external load is essentially the same for different raster angles; refer to Figure 7.7.

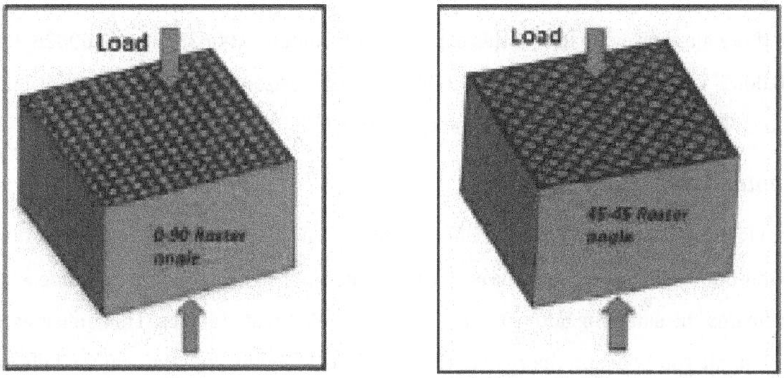

Fig. 7.7: Physical Model Horizontal Build Solid Model with (0°, 90°) and (45°, -45°) Raster Angles [151]

Vetical Build

From the graphs in Figures 7.5 and 7.6, the vertical-build specimens with (0°, 90°) raster angles exhibit about 12% higher yield strength in comparison to the specimens with (45°, -45°) raster angles. The trend remains the same for different air gaps. This indicates that the specimens built with (0°, 90°) raster angle offer more resistance to deformation in comparison to the specimens built with (45°, -45°) raster angle. Figure 7.8 illustrates the (0°, 90°) vs. (45°, -45°) raster angle in vertical-build solid specimens.

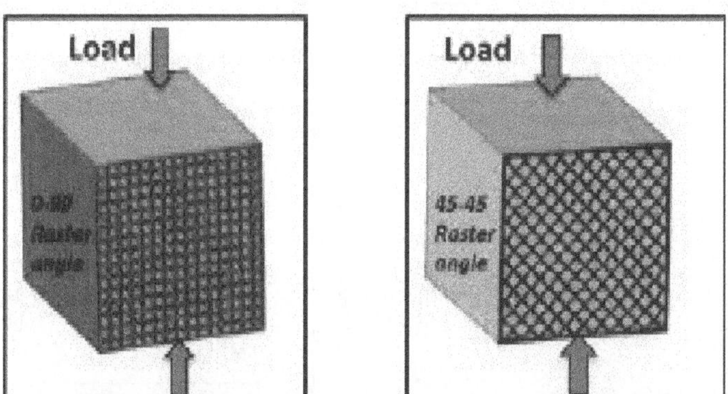

Fig. 7.8: Physical Model Vertical Build Solid Model with (0°, 90°) and (45°, -45°) Raster Angles [151]

7.5.6. Effect of Air Gap

The three air gaps evaluated in this experiment included -0.00635 mm (-0.00025"), -0.0127 mm (-0.0005") and -0.01905 mm (-0.00075") for both horizontal-build and vertical-build solid specimens. The results are shown in Figures 7.9 and 7.10.

Horizontal Build

From Figures 7.9 and 7.10 and the effects table in Table 7.6, it can be seen that in the case of horizontal-build solid specimens, the effect of air gap is not significant with a p-value > 0.05. The trend remains the same for both (0°, 90°) and (45°, -45°) raster angles. This indicates that the variation in air gap between the rasters does not have a significant effect on the failure due to inter-layer sliding in solid specimens.

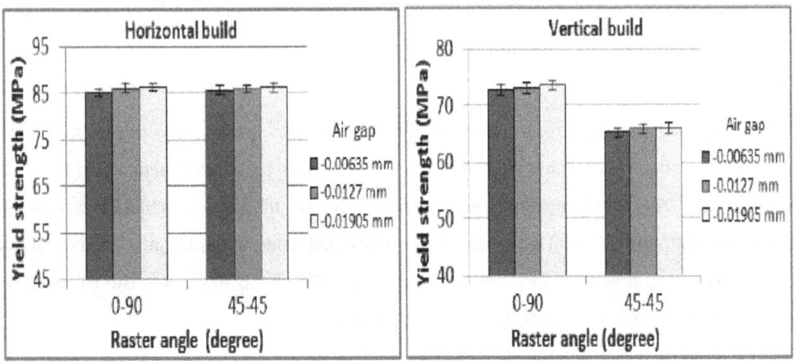

Fig. 7.9: Effect of Air Gap on Yield Strength for Solid Specimens [151]

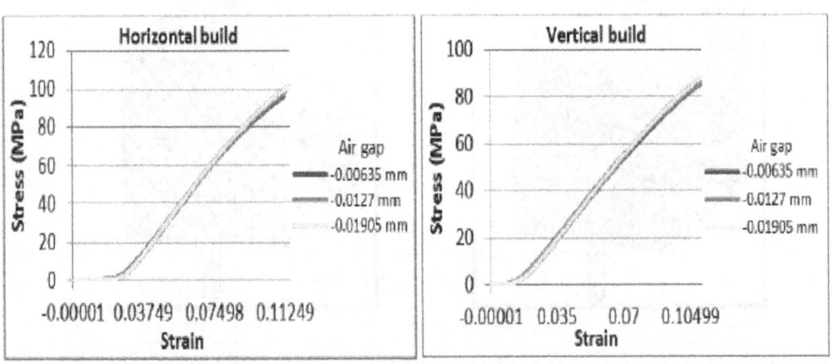

Fig. 7.10: Stress-strain Curves at Different Air Gaps for Horizontal-build and Vertical-build Specimens, Where the Raster Angle was (0°, 90°) [151]

Vertical Build

The effect of air gap in vertical-build is also not significant and the trend remains the same for (0°, 90°) and (45°, -45°) raster angles. As indicated by the effects table, the influence of air gap on the yield strength is not significant (p-value > 0.05), i.e., there is no main effect of air gap for vertical-build. This indicates that the variation in air gap between the rasters also does not have a significant effect on the failure due to buckling in solid specimens.

7.6. Compression Test

Kate et al experimented the compression test based on the ISO-604 standard, where type B specimens were used. The specimens were loaded until the material yielded under the applied compressive force. The different parameter combinations for the different specimens tested did not show the parameters having much influence on the results obtained unlike the tensile test. All the specimens were permanently deformed, with all specimens being compressed to an S-shape [152]. For the flat printed specimens, different layer orientations showed similar trends in compressive strength, despite having small value difference, as shown in Fig. 7.11.

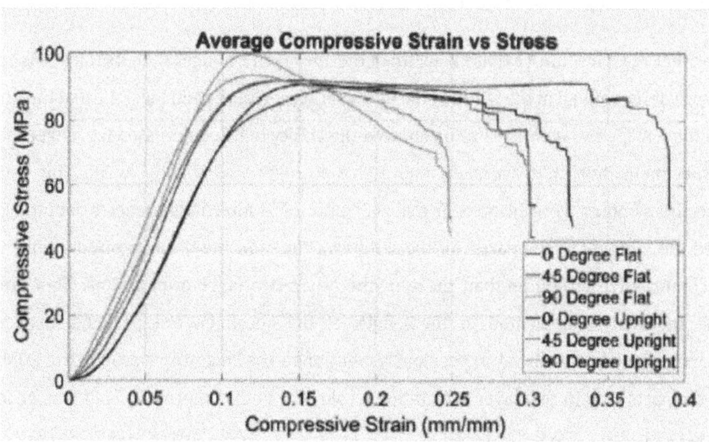

Fig. 7.11: Graph of Average Stress-strain Results for Compression Test [152]

The 90° orientation showed the highest average compressive strength (100.1 MPa) (Table 7.7), though it is not as compressed as the other specimens that are printed in the other two orientations, as can be observed in Fig. 7.11. This could be because the rasters deposited along the width direction of the specimen are parallel to the applied compression load and because FDM parts significantly show an anisotropic property i.e. the resistance of parts in the direction of filament deposition is higher than their resistance in the other two directions.

Table 7.7: Average Results of Compression Extension, Compressive Strength, Compressive Strain and E-modulus [152]

Parameters		**Max. Compressive strength (MPa)**	**Standard deviation of Compressive strength (MPa)**	**Compressive strain at Max compressive strength (mm/mm)**	**Standard deviation of compressive strain (mm/mm)**
Flat (XY)	0^0	93.6	3.92	0.1296	0.012
	$+45^0/-45^0$	94.3	5.21	0.2756	0.122
	90^0	100.1	4.61	0.1161	0.006
Upright (ZX)	0^0	90.9	2.47	0.1556	0.001
	$+45^0/-45^0$	91.9	0.79	0.1354	0.009
	90^0	93.3	1.76	0.1209	0.006

The other reason could be that the width of the specimen is longer than its thickness and the rasters are densely filled in the width of the part. It is also evident from Fig. 7.11 that the samples built in the +45°/- 45°layer orientation showed the ability to be compressed for longer than the other specimens, but their average compressive strength was slightly lower. This could be because the samples were printed in the +45° and - 45° alternating layer orientations that elevated the stiffness of the parts. All the samples printed in the 0° layer orientation showed lower compressive strengths than those in the other two layer orientations. This could be because the rasters deposited in the length of the specimen are perpendicular to the compressive load direction and, as previously mentioned, the anisotropic property of FDM parts makes the parts built in that layer orientation weaker for both compression and tension loads.

Fig. 7.11 also shows the results for specimens printed in the upright build direction with the three different layer orientations. Similar to the flat build direction, the specimen with 90° layer orientation showed the average maximum compressive strength. However, the reason could be different, as the rasters are perpendicular to the load application direction, contrary to the previous case. Thus, the reason could be the dense structure obtained due to the high number of rasters in the length of the specimen that increased the strength of the part. The samples with +45°/- 45° layer orientation built in this direction showed an intermediate performance.

This could be due to the incomplete fill, similar to the tensile specimen, observed at the edges of the samples that reduced its strength, despite the parts being built with alternating layers of +45° and - 45° layer orientations, giving some stiffness to the parts. Samples with 0° layer orientation printed in the upright build direction showed the lowest average maximum compressive strength of 90.9 MPa among the samples built in both build directions, as detailed in Table 7.7. This might be due to the few rasters deposited in the thin thickness of the sample weakening the part.

7.7. Flexural Behavior

Kate et al performed three point flexural test in accordance with ISO 178. The results of the experiment are reported in Fig. 7.12 and Table 7.8. It is very clear from Fig. 7.12 that the samples printed with 0° layer orientation in the edge build direction showed a higher level of performance compared to those printed with the other two layer orientations.

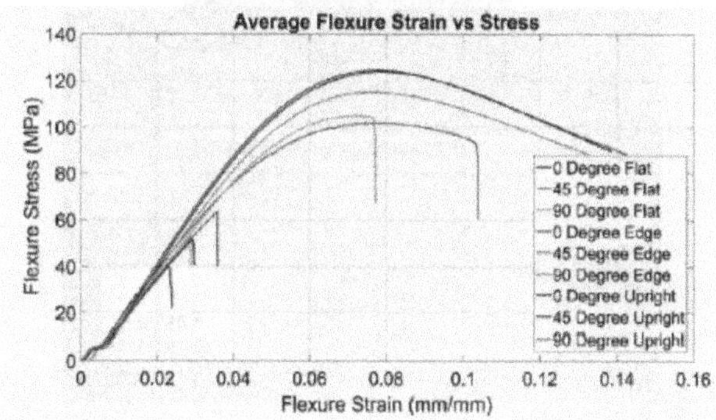

Fig. 7.12: Graph of Average Stress-strain Results for Flexural Test [152]

Table 7.8: Average Results for Three-point Bending Load [152]

Parameters		Max. Flexural strength (MPa)	Standard deviation of Flexural strength (MPa)	Flexural strain at Max Flexural strength (mm/mm)	Standard deviation of Flexural strain (mm/mm)
Flat (XY)	0^0	124.9	0.39	0.0800	1.63×10^{-3}
	$+45^0/-45^0$	102.3	1.31	0.0818	1.55×10^{-3}
	90^0	58.3	5.83	0.0357	5.88×10^{-3}
Edge (XZ)	0^0	123.5	0.32	0.0810	0.90×10^{-4}
	$+45^0/-45^0$	115.4	0.50	0.0782	1.55×10^{-3}
	90^0	106.3	0.84	0.0758	1.05×10^{-3}
Upright (ZX)	0^0	64.2	2.99	0.0403	1.40×10^{-3}
	$+45^0/-45^0$	47.5	4.85	0.0310	3.38×10^{-3}
	90^0	64.9	5.80	0.0396	5.85×10^{-3}

This could be because the samples are built with a dense structure, like that of the tensile specimens (Figure 7.13), elevating the strength of the samples. For fracture to occur, the load needs to separate and break several layers, contours and rasters. The other reason could be the fact that the deposited filaments are perpendicular to the applied downward force and it is very difficult to break the filaments apart. The samples with +45°/- 45°layer orientation printed in the edge build direction also showed higher average maximum flexural strength of 115.4 MPa, compared to 105.3 MPa for the sample with 90° layer orientation printed in this build direction, as given in Table 7.7.

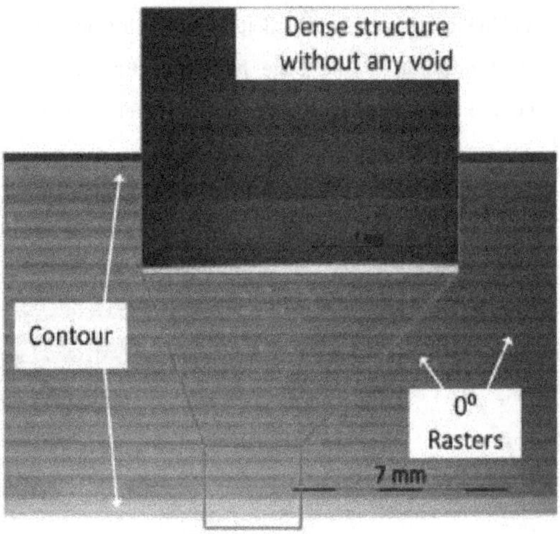

Fig. 7.13: Light Microscope Image Showing Dense Sample Built in Flat Direction with 0° Layer Orientation [152]

These results are comparable the values reported on the material datasheet [153]. This could be because building the part in the edge direction gives a higher number of contours and the rasters are deposited in an alternating +45° and–45° layer orientation fashion, giving the part greater strength, despite incomplete fills, similar to the tensile specimens as illustrated in figure 7.14.

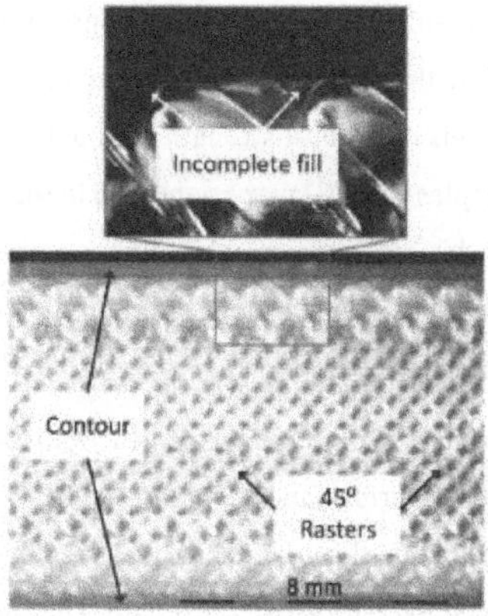

Fig. 7.14 Light microscope image showing samples with flaws when built in flat direction with +45°/- 45° layer orientation [152]

The samples printed with 90° layer orientation in the edge build direction showed inferior performance, compared to the other two layer orientations and they fractured just after reaching the maximum strength. This could be because the filaments deposited show incomplete fills at the edges of the samples where the rasters meet the contour, as shown in Figure 7.15. This can cause their performance to deteriorate, as well as premature failure, regardless of the higher number of contours that can add strength to the part. It could also be due to the deposition direction of the filaments being parallel to the flexural load applied, making it easy for the force to separate the filaments and cause fracturing at their interface.

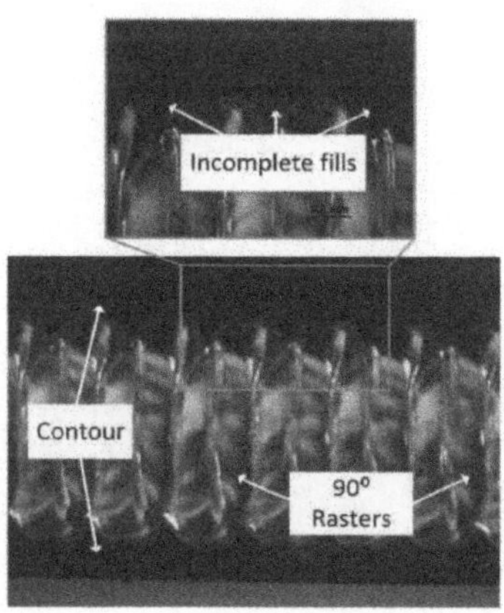

Fig. 7.15: Light Microscope Image of Incomplete Fills for the Sample with the 90° Layer Orientation Printed in the Edge Direction [152]

Figure 7.12 clearly shows results for flat build directions for all the layer orientations. It is very clear from the figure that the samples with 0° layer orientation built in this direction showed exceptionally high levels of flexural strength of about 125 MPa. This could be for the same reason as that of the samples with 0° layer orientation in the edge build direction; the very dense structure achieved for this parameter combination is very strong. It is also because the applied load is perpendicular to the filament deposition direction, which makes the part very tough and able to sustain the applied load.

The samples built with parameter combinations of +45°/- 45° layer orientation and flat build direction exhibited intermediate performance compared to the other two-layer orientations. This could be for the very reason that the rasters are deposited in an alternating +45° and - 45° layer orientation fashion, giving the part greater strength. The samples printed with 90° layer orientation in the flat build direction showed exceptionally poor performance. The reason for this could also be that the direction of the filaments is parallel to the flexural load applied, making it easy for the force to tear the filaments apart. The sample printed in the upright direction showed very poor performance for all the three layer orientations. This could be because the parts are built with the stabilizing wall, which not only makes poor adhesion between the consecutive layers but also leaves some traces at the connection with the part, making the part very rough.

CHAPTER 8

8. Research Trends in Polymer Composites

8.1. Introduction

Fused Deposition Modeling (FDM) is a widely used additive manufacturing technology for fabrication of complex geometric parts using thermoplastic polymers. The quality issues and inferior properties of fabricated parts limited this process to manufacture parts for industrial level applications. Reinforcing the polymer with nanoparticles, short fibers or continuous fibers improve mechanical, thermal and electrical properties compared to the neat polymer. The success of expanding this technique to industrial applications depends on the preparation of printable composite feedstock filament and printing without defects.

The challenges involved in the preparation of composite feedstock filaments and printing issues during the printing of nano composites, short and continuous fiber composites are highly challenging one. Initially, FDM printed parts have been used for low load prototypes such as domestic toy applications. With technological advancements, the process is extended nowadays to print parts for real time industrial applications. Large scale geometries of several meters of length and width are recently printed at very high material output rates of more than 200 kg/h [154].

To improve the properties of the FDM printed parts, literature works focused on developing composite material systems by reinforcing different fillers to the base polymer and printing of these composites Parts made using these composites found to exhibit higher mechanical, thermal and electrical properties compared to the unreinforced polymer printed parts. Driven by new applications and challenges, the FDM technology keeps growing in producing components made of new composite material systems [155]. The properties of composite materials are highly sensitive to the reinforcement type, i.e., particle, short fiber or continuous fiber. Also, the reinforcement size, shape and distribution affect the properties. Generally, in the case of nano particles, more than 8–10 wt% may decrease the properties due to agglomerates formation. Short fiber composites show a decrement in properties at loadings higher than 30 wt%.

During the FDM printing of short fiber or particle reinforced composites, the orientation of fibers in the extruding nozzle and in the print layer depends on the flow field in the liquefier and in the nozzle [156]. The reinforcement size alters the melt viscosity, which effects the flow field.

The nozzle may act as a filter for higher particle sizes, which causes the nozzle blockage after few prints.

Few thermal and electrical properties of 3D printed parts using short-fiber composites have been improved. However, the mechanical performance of printed parts with particle or short fiber composites are still inferior to the parts made using other short fiber composite processing techniques such as compression molding. The matrix material (in the filament form) has been printed in one stage and reinforcement material in another stage as shown in figure 8.1.

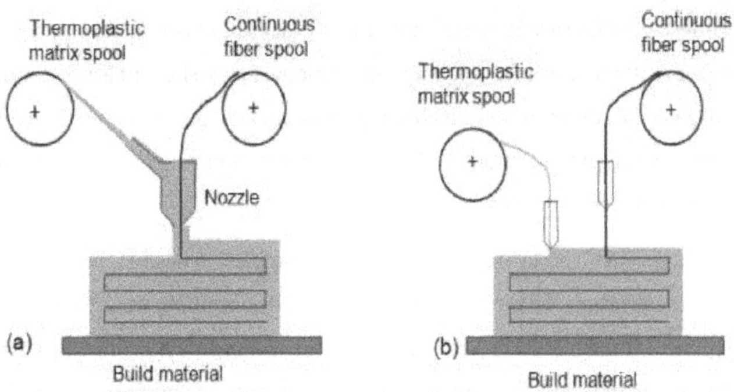

Fig. 8.1: Schematic of FDM Process Using Continuous Fiber Reinforcement a) In-nozzle Impregnation b) Printing with Two Separate Nozzle [157]

The mechanical performance of continuous fiber reinforced composites is higher compared to short fiber composites. The properties of continuous fiber reinforced composites can be controlled by volume fraction of fibers. Few works reported about the printing of continuous fiber reinforced composites using two separate print heads.

8.2. FDM of Thermoplastic Composites

Major issues during the FDM printing are the presence of void content in the microstructure, dimensional inaccuracy and the anisotropy of the printed parts. They arise due to the weak intra layer bonding between beads in a single layer and weak inter layer bonding between different layers in the thickness direction. Therefore, the strength of printed parts is limited. All research work in the area of layer wise additive manufacturing mainly focus on overcoming one or two of these problems. Even though polymer composites exhibit higher mechanical properties than polymers, properties of composite printed parts are still inferior. This is due to presence of weak reinforcement/polymer interface in case of continuous fiber reinforced composites and poor interface stress transfer in case of short fiber or particle reinforced composites.

Particularly, the processing route of nano composites has major effect on the interface strength. In this review, discussion is mainly focused on void content, anisotropy and interface strength of the parts made via FDM of thermoplastic composites.

8.3. FDM of ABS Composites

ABS is a widely used thermoplastic material in FDM because it has sufficient melt fluidity, desirable stiffness and strength. The tensile, compressive and flexure strength of FDM parts made of ABS are low compared to the parts fabricated by injection molding. For instance, FDM fabricated ABS has only 74% tensile strength to that of injection molded ABS [158]. This is due to high material compaction in the injection molding process. Also, in FDM, layer by layer addition induces anisotropy in the built part. Researches have conducted various experiments to investigate the effect of different process parameters on the anisotropy and strength of ABS printed parts. The negative air gap and raster orientation has a major influence on mechanical properties. Dawoud et al. reported that providing a negative raster gap will result in high packing of the material and hence properties are improved [159]. To improve mechanical properties and mitigate the print orientation induced anisotropy, several studies have been performed to print ABS composites. Zhong et al. conducted experiments to investigate the process ability of short glass fibers reinforced ABS matrix composites. The addition of glass fibers decreases the ductility. The filament could not be wounded into the spool after extrusion due to brittle nature of the composite. Addition of plasticizer makes the composite ductile and filaments of constant diameter can be prepared [160].

8.3.1. Carbon based Reinforcement

Carbon based reinforcements improve the mechanical and electrical properties of polymers. Shofner et al. developed ABS matrix composites by reinforcing vapour grown carbon fibres (VGCF) and single-walled carbon nanotubes (SWNT) [161]. Tensile tests have been conducted on three different specimens: thin sheets of composites, extruded composite filaments and on the parts made of extrusion freeform fabrication. Both of these nanofillers dispersed homogeneously in ABS and the composites showed an increase in strength and modulus.

During the filament extrusion, reinforcements are aligned along the extrusion direction. The extruded fiber showed an increase in strength and modulus compared to the composites. Despite preferred orientation in the filaments, the strength of the printed parts showed decrease in strength and modulus due to voids. Compared with unfilled ABS printed parts, nearly 40% and 60% increase in tensile strength and tensile modulus were obtained at nanofiber loading of

10 wt% [162]. Ning et al. reinforced short carbon fibers of 7.2μm diameter and 100μm length in the ABS [163].

The improvement in tensile strength and elastic modulus of the printed parts is observed at 5% and 7.5%, respectively. As shown in Figure 8.2a and 8.2b, three types of voids have been reported in this study. Pore type 1 is due to gas evolution. Pore type 2 is the inter layer void and pore type 3 is the void at the interface between fiber and matrix with the bead. It is also reported that the increase in the length of carbon fibers increases the strength and modulus but decreases the toughness and ductility.

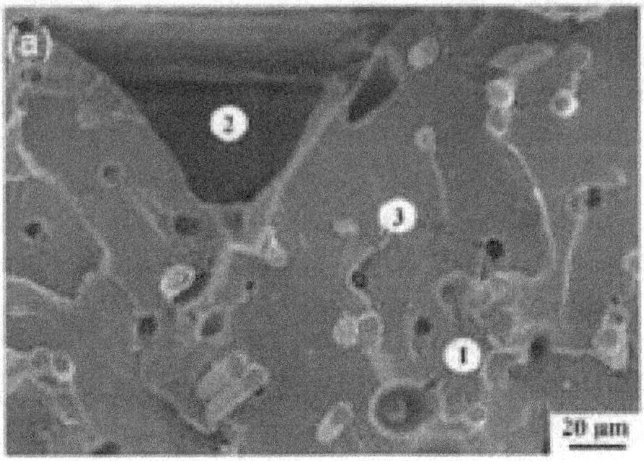

Fig. 8.2a: The Presence of Voids in the FDM Printed Composite Parts a) ABS+10 wt. % Short Carbon Fibers of 150 μm Length [163]

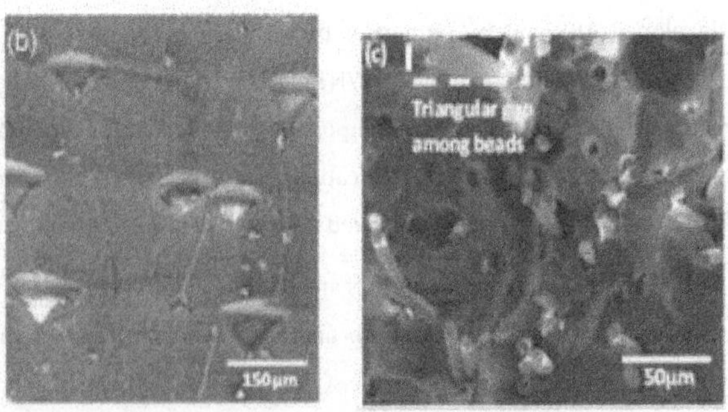

Fig. 8.2b: The Presence of Voids in the FDM Printed Composite Parts b) Neat ABS c) ABS+10 wt % Short Carbon Fibers of 0.35 mm Length [164]

An important observation has been reported about the voids in the printed parts. Triangular shaped voids are formed between the beads in the printed parts and the void size is reduced compared to the pure ABS printed parts. However, with the addition of short fibers, additional voids have also been observed within the beads due to the poor interface between carbon fibers and ABS. Up to 30 wt % of carbon fiber loading, a great improvement in strength and modulus has been observed, but nozzle clogging was observed for higher loadings.

Zhang et al. processed FDM parts using carbon nanotube reinforced ABS and carbon fiber reinforced ABS composites. It is observed that the strength and modulus is of printed parts is higher for these two composite material systems, though the void content in the printed parts is higher compared to pure ABS printed parts. Also, the carbon nanotube reinforced ABS printed parts showed less void content than the carbon fiber reinforced ABS [165].

Printing at a less layer thickness of 0.8 mm, low printing speed of 60 mm/s and 0 ∘ degree raster orientation showed the highest properties. Love et al. investigated anisotropy of ABS + short CF composite 3D printing [166]. It is reported that the composite print parts showed anisotropy. The tensile properties of specimen printed along longitudinal direction are higher than the properties of the specimen built vertically. Also, it is reported that the addition of short carbon fibers increases thermal conductivity and increases thermal expansion coefficient which are key to the dimensional stability of the printed parts. Weng et al. prepared nano composites by reinforcing organically modified montmorillonite (OMMT) by melt intercalation [167]. The addition of 5 wt% of OMMT significantly improved both tensile and flexural strength and modulus of the printed parts.

8.3.2. Graphene based Reinforcement

The new generation material graphene attracted most attention recently, which will improve the properties when reinforced with polymers. However, the dispersion of graphene may not be good in majority of polymers. The graphene oxide (GO) has better dispersion ability, however, the electrical properties will be decreased because of the presence of sp^3 bonds in GO. Reducing GO in reductants such as Hydrazine restores sp^2 bonds and improves the electrical properties. Wei et al. processed graphene oxide (GO) reinforced ABS composites using the solvent method for different weight fractions. The two solutions, GO sheets in NMP and ABS in NMP were mixed using homogenizer and this mixture was later reduced in Hydrazine. Filaments of Graphene-ABS composites are prepared using a single screw extruder and used for FDM printing [168].

It is demonstrated that the graphene nanocomposites with up to 5.4 wt % graphene are 3D printable and beyond this graphene loading, the nozzle blockage occurs. The printed parts showed enhanced electrical conductivity properties than pure ABS parts. The selection of a proper solvent for the components to dissolve initially and remove the solvent from the composite is challenging. Dul et al. processed graphene nanoplatelets (GNP)-ABS composites using melt compounding (dry mixing) [169]. The GNPs were added to the ABS melt and mixed using the counter-rotating extruder. It is observed that the elastic modulus increases with graphene loading, but tensile strength and elongation to breakage decreases. Also, at filler loadings higher than 4 wt%, nozzle block is observed during printing. The size of the reinforcement is the key issue as higher size particles may block the nozzle. Aumnate et al. compared the GO/graphene composites prepared using both the solvent and melt compounding methods. It is reported that the size of fillers are in the order of 10μm and 3μm for dry mixing and solvent mixing routes, respectively. The GO/ABS filament made from the dry mixing failed to print due to the agglomeration of GO and die clogging [170].

8.4. FDM of PLA Composites

PLA is another common thermoplastic material used in FDM, which has wide applications in the medical industry due to its biodegradable/ biocompatible properties. It has a lower ductility than ABS but has higher strength. Similar to FDM of ABS composites, an increase in void content and the anisotropy is observed during the printing of PLA composites. For example, Ferreira et al. processed FDM parts using PLA and chopped short carbon fiber (60μm length) reinforced PLA. The fibers were observed to be aligned along the filament's length during extrusion and are aligned along the print direction during printing. It is observed that tensile and shear modulus of the composite printed parts are higher than about 100% compared to pure PLA printed parts.

The failure strain of PLA + carbon fiber composite is lesser than PLA because the composite became more brittle after adding short carbon fibers. Also, there is not much change in tensile strength is observed with the addition of carbon fibers. The reinforcement of carbon fibers in PLA results in a weak bonding interface due to the poor adhesion of reinforcements.

If the length of reinforcement is increased, good adhesion may be observed because of the increase in the external surface area. While increasing fiber length enhances the adhesion in ABS composites, PLA composites show poor adhesion for longer fiber lengths. Investigations have been carried out to find the effect of other carbon derivatives such as carbon nanotubes and graphene on PLA.

Plymill et al. prepared nano composites by reinforcing graphene and multi-walled carbon nano tubes (MWCNT) individually using twin screw extruder, which provides a high shear rate mixing [171]. The resulting composite in pelletized form is fed into a single screw extruder for preparing the FDM feedstock filament. Filaments of constant diameter have been prepared by proper control of the cooling rate at the nozzle exit. An increase in the tensile strength of the printed parts has been observed at 0.2 wt% loading of graphene and 0.1 wt% loading of MWCNT.

Bustillos et al. printed FDM parts from the commercially available conductive PLA-graphene filament from BlackMagic © [172]. Figure 8.3 compares the microstructure of the PLA and PLA-graphene printed parts. The inter layer adhesion is much stronger in PLA and voids are observed between the layers during the printing of composite.

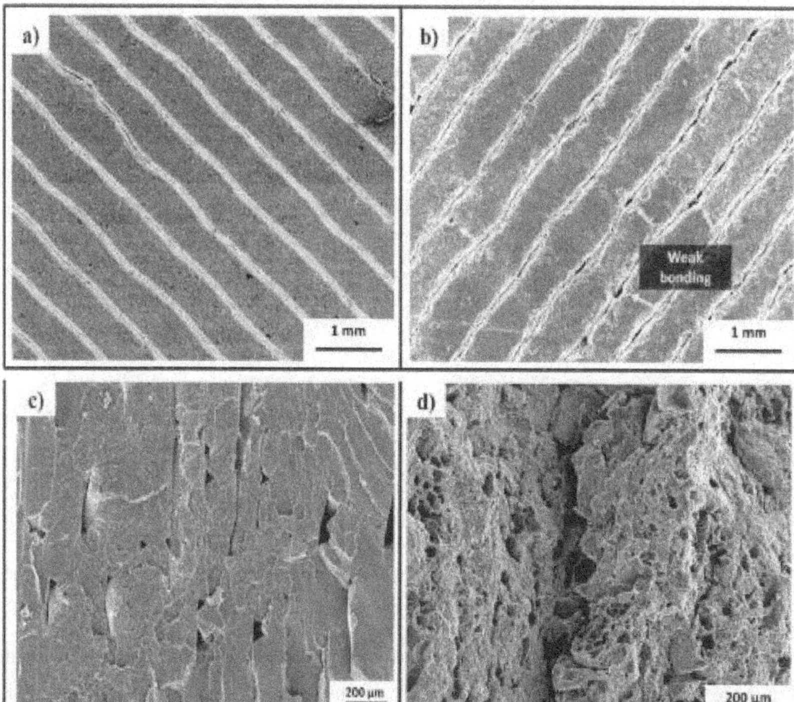

Fig. 8.3: Microstructure of 3D Printed (a) PLA and (b) PLA-graphene Parts. Fractured Cross-sectional. Microstructure of 3D Printed (c) PLA and (d) PLA-graphene [172]

It is reported that the addition of graphene nanoplatelets increases tensile and flexural strength but decreases the impact strength. The anisotropy of the printed samples has been investigated by printing the specimens along flat, on-edge and upright positions. The tensile, flexural and impact strength has been decreased by almost 30% for an upright position. Also, it is reported that the presence of graphene nanoplatelets increases the surface texture without effecting the geometrical dimensional accuracy.

The weak strength along the transverse direction of the printed parts is attributed to the poor inter layer adhesion of polymers. The adhesion is stronger if the polymer surface is heated above its sintering temperature. However, during the FDM process, the cooling is faster because of the short times involved, and hence the inter layer adhesion is weaker. To avoid this problem, trails have been made to increase the heating time using air nozzles. However, this process resulted in the high dimensional inaccuracy of printed parts. An interesting study has been presented by Sweeney et al. to increase the interlayer strength by local heating of polymer interfaces [173]. In this work, the PLA filament is coated with CNTs before printing as shown in figure 8.4.

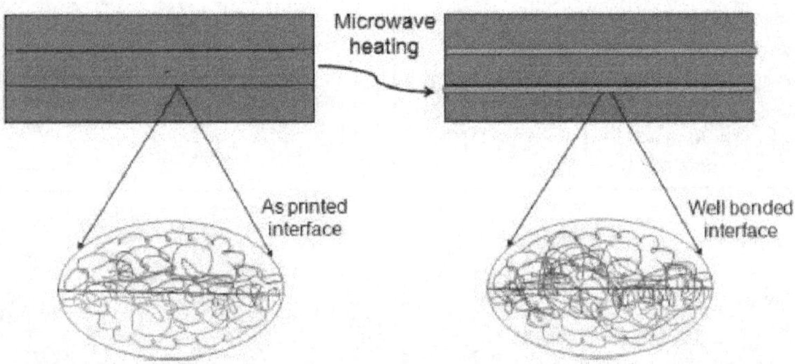

Fig. 8.4: The Interface is Locally Heated to Allow for Polymer Diffusion and Increased Fracture Strength [173]

During the printing process, the localized heating at nano scale using microwaves causes the surfaces to melt and promotes inter layer adhesion. The study reported an increase in fracture strength by 275% with this local welding by microwave heating. Further, Ivanpv et al. Reported that the addition of bifillers (both CNT and Graphene) increase electrical properties and reduce agglomeration. PLA printed parts also has excellent bio compatibility as explained later in the section on applications of FDM [174].

In continuous fiber reinforced composites, the poor adhesion between matrix and fiber interface deteriorates mechanical properties. Matsuzaki et al. processed continuous carbon fiber reinforced thermoplastic (CCFRTP) composites using continuous carbon fiber tows and jute twisted yarns as reinforcements in PLA and these composites have been 3D printed. The study reported that the tensile strength and modulus of CCFRTP composite printed parts are higher than the part properties using other composite printing techniques such as selective laser sintering (SLS). These mechanical properties are also higher than the FDM printed parts using short carbon fiber reinforced composites.

It is also reported that mechanical properties of these parts can be further increased by increasing the volume fraction of continuous fibers. However, poor adhesion between the matrix and fiber has been observed. In general, the surface treatment of fibers enhances the mechanical properties of composites. Li et al. processed CCFRTP printed parts after applying the PLA compatible sizing agent such as methylene dichloride [175].

As shown in Figure 8.5, the fiber-matrix interface adhesion is improved in the printed parts. The resulting 3D printed parts showed an improved tensile and flexural strength compared to PLA and unmodified composite. Tian et al. studied FDM process parameters such as liquefier temperature, layer thickness on the interface and performance of the continuous fiber 3D printed structures [176]. The study also showed the capability of the FDM process for printing large curvature products using CCFRTP. In another study, Tian et al. recycled the FDM printed continuous fiber reinforced PLA composites and the resulting materials were again 3D printed [177]. The parts made of recycled composite showed a better tensile and flexural properties because the residual impregnation of PLA improved the fibre-matrix adhesion.

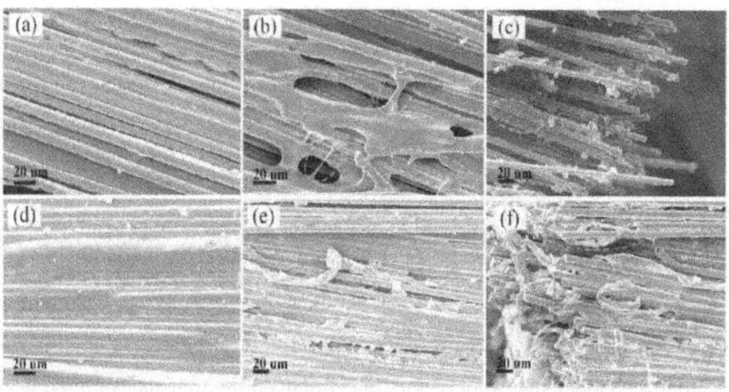

Fig. 8.5 a) Thermoplastic PLA Filament Coated with CNT-rich Layer b) After the Microwave Heating, The Interface is Locally Heated to Allow for Polymer Diffusion and Increased Fracture Strength [175]

8.5. FDM of Polyamide

Nylon has properties similar to that of commercial ABS and can be processed into feedstock filament at a low cost. Particle reinforced nylon composites have been used to print molds for tooling applications. Masood et al. manufactured injection molding inserts through the FDM of nylon + iron metal powder composites. The iron metal powder has good mechanical and thermal properties and a high amount of mixing and surface binding with polymers. Small quantities of plasticizer and surfactants are added to improve the flow, dispersion and toughness properties of the composite. The land length of the die was increased from 5 to 10 mm to avoid a phenomenon known as die swell, which is primarily due to the elastic recovery of the metal-polymer composite in the die. The processed injection molding inserts have desirable mechanical properties. It is reported that the size and volume fraction of the filler has a significant influence on the mechanical properties. Large size fillers increase tensile modulus and small fillers increase the tensile elongation.

Singh et al. processed Al2O3 reinforced nylon matrix composites to develop wear resistant materials for grinding applications with special applications in the dentistry. The amount of filler and its size are chosen appropriately, such that the melt flow index (MFI) of the processed composite is equal to MFI of ABS. The wear tests on the FDM prepared parts using these composites revealed the wear resistance of this composite higher than the Al2O3 – ABS composite. Further, the authors performed experiments to investigate the effect of particle size. Nylon composite FDM feed stock filaments have been prepared using single particles and a mixture of two and three particles of different diameters. The particle size of more than 150 μm may cause nozzle clogging and the size of all fillers used is less than that in all cases. MFI of all mixtures is near to commercial ABS material (2.41 g/10 min) [178].

Abdullah et al. processed zirconium oxide filled nylon composites to use in FDM for printing patient specific implants. At higher loadings and for the particle size of 400μm, agglomeration of particles occurred in the composite and this caused the nozzle blocking during 3D printing by a 0.4 mm nozzle. Due to this, the properties of the printed parts are reduced compared to the properties of pure polyamide printed parts [179]. Zhu et al. prepared composites of polyamide 12 with 6 wt% of graphene nanoplatelets (GNP) in pellet form by melt compounding using twin-screw extruder. FDM filaments are prepared by single screw extruder from these pellets. It is reported that the 6 wt% of graphene loading enhanced the thermal conductivity and elastic modulus of 3D printed parts by 50% and 7%, respectively.

The scanning electron microscope (SEM) images of the failure of 3D printed samples showed that the GNPs are highly oriented along the print direction during the nozzle extrusion. The authors did not report the nozzle blocking issues at this high loading of GNPs. The addition of GNPs increases the melt viscosity, which effects the coalescence [180]. Currently, nylon is mostly used as the matrix material for commercially available continuous fiber reinforced printers. To increase the adhesion of fibers to the nylon matrix, fibers are also pre-impregnated with nylon. For instance, nylon impregnated glass and carbon fibers are commercially available. The in-situ fusion in the in-nozzle impegration technique can be useful for controlling the fiber volume fraction, however weak matrix and fiber interface remains a challenge.

The printing using the pre-impregnated continuous fiber may have a better interface, but the amount of volume fraction of fibers is limited. Goh et al. processed continuous fiber-reinforced thermoplastic carbon and glass composites (CFRTP and GFRTP) using FFF [84]. Nylon matrix pre-impregnated fiber filaments are used as feed stock. The authors carried out tensile, flexural, and indentation tests and the failure mechanisms in each mode have been analyzed. As shown in Figure 8.6, tensile rupture perpendicular to loading direction and shear damage along the fiber direction is observed as the main failure mechanism in tensile tests. Delamination and debonding between the adjacent fiber filaments have also been observed. In the bending tests, cracks initiated at the peak compression place in CFRTP and propagated towards the neutral plane due to lower compressive strength of carbon fibers.

Catastrophic failure occurs when the tensile side stress reaches its peak value. No crack observed in GFRTP specimens, rather buckling of fibers is observed due to shear kinking and matrix failure. In the indentation tests, debonding between adjacent filaments, delamination and fiber breakage has been observed as the main failure mechanisms. Dickson et al. processed continuous fiber reinforced composites using two separate print heads in which nylon was printed in one stage and reinforcement in another stage. Initially, the fiber bundle will be printed with a concentric or isotropic pattern, and then the nylon matrix will be printed in the gaps before the next layer of fiber printing.

Continuous glass, carbon and Kevlar individual fiber reinforced nylon matrix composites have been printed. It is reported that the isotropic pattern yields excellent properties in tension and concentric pattern yields better properties on bending. It is also reported that with an increase in fiber volume fraction, the strength of continuous fiber reinforced composites increases. For example, the tensile strength of 33 vol % of glass fibers printed in a concentric pattern matches the strength of aerospace grade aluminum [182].

This example highlights the potential of FDM to print parts for real time applications with control of the fiber volume fraction. However, as the fiber content increases, the void percentage also increases thus limiting the maximum fiber volume fraction.

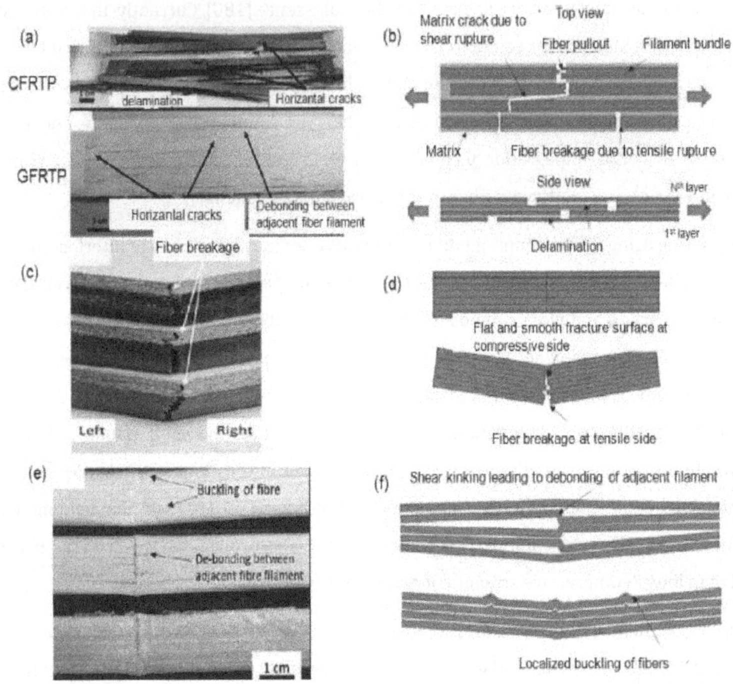

Fig. 8.6: Illustration of the Failure Mechanism of 3D Printed Continuous Fiber Reinforced Composites, Tensile Behavior: a) and b), Flexure Behavior of Carbon FRTP: c) and d) Flexure Behavior of Glass FRTP: e) and f), [181]

8.6. FDM of Polypropylene

Polypropylene (PP) is a semi-crystalline polymer that is extensively used in packaging, household furniture, medical and automobile applications. It is a promising thermoplastic material with its high impact strength and low cost. Parts made of it possess dimensional inaccuracies because of the high thermal expansion and warping nature. Carnerio et al. demonstrated that the semi crystalline nature of PP causes shrinkage of the printed parts leading to distortion and decoupling from the print bed, as shown in figure 8.7 a [183].

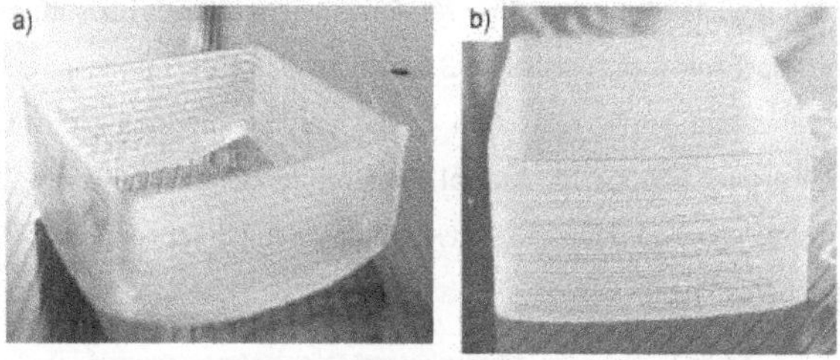

Fig. 8.7: FDM Printed Polypropylene a) With Distortion and Decoupling From the Bed, b) Reduced Distortion and Improved Adhesion to the Bed at Optimal Printing Parameters [183]

Maintaining the nozzle temperature at 165 $^\circ$C and printing the initial layer (that is in contact with unidirectional orientation drastically reduced the warpage as illustrated in figure. 8.7 b). It is to be noted that the adhesion of polypropylene to the regular print surfaces is very poor. Changing the print surface to an adequately cleaned polypropylene plate promotes the adhesion. Other studies on FDM printing using PP reported that the shrinkage and warpage could be reduced by adding fillers and controlling the build chamber temperature.

Spoerk et al. [184] processed PP composites with expanded-perlite fillers and found that the addition of small size fillers results in the reduction of shrinkage and warpage. The small sized fillers enhance the filler-matrix adhesion, reduce the polymer chain orientation and constrain the polymer from shrinking. An increase in the build chamber temperature alters the morphology of the crystal growth. For example, maintaining the chamber temperature at 55 $^\circ$C resulted in spherulites formation, which causes a decrease in the warpage. However, the addition of fillers decreases the impact strength.

The effect of build chamber temperature on the crystal growth and mechanical property enhancement needs to be further investigated. The FDM feed stock filament prepared from glass-fiber PP composites maintained the minimum critical length after the extrusion. However, the properties of the 3D printed glass PP composites are still lesser compared to the parts made of compression molding. Composites are made by reinforcing harakeke, hemp fibers and gypsum powders in the recycled PP. The composites are initially prepared by blending PP and reinforcements in a twin-screw extruder. For the natural fiber reinforcements, beyond 30 wt% of hemp and harakeke fibers, the dimensional accuracy of the filament is not properly maintained and a 'Sharkskin' effect has been observed on filaments. However, there is no such effect was observed with the addition of gypsum powder.

It is reported that while the natural fiber reinforcement increases the stiffness and strength of filaments, the mechanical properties of 3D printed parts decreased possibly due to the formation and voids and stress relaxation of the polymer occurring during 3D printing. However, the 3D printed parts with natural fiber reinforcement showed a decrease in the distortion. For instance, the addition of 30 wt%. Natural hemp fiber reduces the part distortion significantly, as shown in figure 8.8.

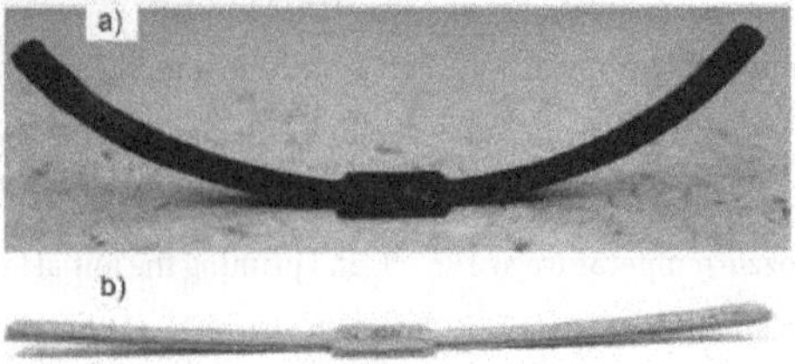

Fig. 8.8: a) Shrinkage of the FDM Printed Polypropylene Part; b) Reduction in Shrinkage of PP with 30 wt%. Natural Fiber Hemp [185]

8.7. FDM of Polyethylene

Polyethylene (PE) is a commonly used thermoplastic polymer in daily life. PE polymers are categorized as low density polyethylene (LDPE), high density polyethylene (HDPE), ultra high molecular weight polyethylene (UHMWPE), etc. PE polymers have excellent recyclability and are potential materials for sustainable manufacturing. The processing of PE via FDM adds a new dimension to the additive manufacturing technology due to recyclability of parts. However, PE has the highest volume shrinkage rate upon solidification, a high degree of warpage and does not easily stick to the print platform. Similar to PP the adhesion to the print platform can be enhanced and the warpage of the printed part can be decreased by selecting optimal print and process parameters. Schirmeister et al. suggested that using SEBS (styrene-block-ethene-co-butene-block-styrene) as a support material showed better adhesion for HDPE printing. The part can be easily detached from the built platform with excellent surface quality.

The same study also reported that providing a gradient in the filling amount can greatly reduce the void content in the printed parts. For example, filling 100% in the first layer and 107% in the sixth layer significantly reduced the void content, refer figure 8.9. Because of the reduction in void content, the mechanical properties of the printed parts with this optimal print parameters are almost equal to the properties of the injection molded (IM) HDPE. The study reported that the elongation to break is less when compared to IM specimens.

However, this property can also be increased by optimizing the process parameters. Additions of high performance filler materials to PE and printing of these composite materials can extend the applicability of FDM. Bedi et al. processed PE/Al2O3 composites with different Al2O3 particle size [187]. The thermoplastic granules collected from house hold waste like plastic films, package bags etc. are blended with Al2O3 particles and the composite filaments are prepared using twin screw extruder. The MFI is decreased for the composite compared to PE.

The addition of glass powder reinforcement improved the wear resistance by 50% and the printed parts are potential candidates for low-duty frictional applications. Tarres et al. processed biocomposites using thermomechanical pulp (TMP) as the reinforcement in HDPE [188]. The study reported that the printing of neat HDPE failed due to improper adhesion. The adhesion is improved for the composite by adding the compatibilizer such as maleic anhydride functionalized polyethylene. The mechanical properties of the printed composites are improved. Polyethylene terephthalate glycol copolymer (PETG) is used in food and medical applications. It has high chemical and impact resistance along with excellent recyclability, biocompatibility. The crystallinity of this polymer is very low (only up to 3%). Therefore, it is very much suitable for extrusion printing. However, parts printed with PETG contain high porosity.

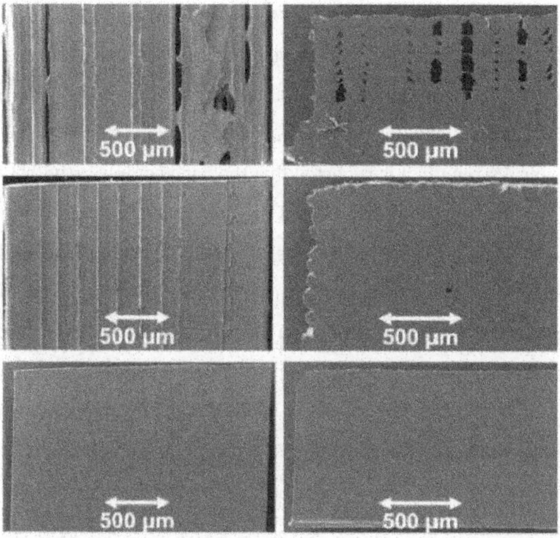

Fig. 8.9: SEM Images of the surface (Left) and Cross Sectional Area (Right) of HDPE Specimens Obtained by FFF (1–4) and by Injection Molding, Respectively (5–6). In Contrast to Regular Printing (1 and 2) FFF with Gradually Adjusted Filling Amount (3–4) is Clearly Superior in Terms Surface Quality and Preventing Defect Formation [186]

8.8. FDM of Polyether Ether Ketone (PEEK)

Poly-ether-ether-ketone (PEEK) is an interesting semi-crystalline high performance thermoplastic polymer with a molecular structure presenting aromatic groups and both ether and ketone group in the backbone. PEEK has superior properties (Tg between 143 and 148 ∘ C, Tm between 330 ∘C and 340 ∘C and service temperatures up to 260 ∘ C). The use of PEEK is increasing in applications ranging from biomedical to high temperature. Filament extrusion and further printing of PEEK is challenging because of the high melting temperature and high expansion upon melting. Kishore et al. reinforced short carbon fibers in PEEK and conducted rheological tests [189]. It is reported that the viscosity of the melt is lower than ABS composites showing the promising nature of processing the PEEK composites into feedstock filaments. Berretta et al. synthesized PEEK and CNT composite specimens with 1 wt% and 5 wt% carbon nanotubes (CNT). The filaments became brittle with the addition of CNT.

The presence of pores was observed in the single layer FDM samples of PEEK and it's composite. The addition of CNT up to 1 wt%. Increase the ultimate tensile strength. However, the UTS is decreased at 5 wt% of CNTs which may be attributed to pores and agglomerates during printing. Also, tensile samples for short beam shear test were fabricated to find layer to layer bonding. Shear strength of PEEK and 1 wt% CNT PEEK were found to be same, whereas 5 wt% CNT PEEK had a significant decrease. As suggested, surface characteristics (surface tension, wettability and adhesion when molten) are different for 5% than 1% [190]. The thermal and electrical properties of PEEK composites are enhanced by adding both MWCNT and GNP with a reasonable mechanical strength. It is reported that the nano composite with 4 wt %. CNT and 3 wt% GNP shows the electrical conductivity of 10 S/m. However, the parts made of FDM using this composition showed a decrease in electrical conductivity due to voids in 3D printed parts. Combining the advantages of continuous fiber reinforcement in high strength PEEK and local heating to increase the inter layer adhesion can dramatically increase the properties of the 3D printed parts. To decrease the void formation and increase the inter layer adhesion, techniques such as infrared lamp heating, hot air nozzle preheating and plasma radiation have been used to print low temperature amorphous polymers. However, these techniques will lead to large volume shrinkage.

CHAPTER 9

9. Recent Trends in Fused Deposition Modeling of Metals

9.1. Introduction

Metal extrusion in additive manufacturing is a fairly new process. Similar to the wildly popular plastic-based FDM process, filament is heated and drawn through a nozzle and then deposited layer-by-layer. This filament is a combination of thermoplastic material and metallic particles. The nozzle moves in the x and y axes across the part for a given layer. The build platform then lowers to make room for new layers. After the part is complete, it is placed into a sintering furnace to burn out the remaining plastic and sinter the metal particles together. Extrusion-based additive manufacturing has been widely used for plastics and polymers, but only recently has developed to create metal parts. While metal extrusion is similar to all other 3D printing processes by building its parts layer by layer, it varies due to the material being deposited through a nozzle under constant pressure and in a continuous stream. This pressure must be kept steady and at a constant speed to enable accurate results. Unfortunately, the actual application of the technology tends to leave small voids and bubbles in the part resulting from the melted mixture leaving the nozzle as it's applied in different directions. This has a significantly negative impact on final part properties and porosity. It is because of these final part limitations that manufacturers of FDM-based metal machines rarely advertise any production part applications [191].

9.2. Disadvantages of Metal Additive Manufacturing

- Limited production potential
- High binder content makes sintering difficult
- Significant part shrinkage in furnace
- Supports may be required
- Parts are lower in density

9.3. Metal Extrusion Process

Metal Extrusion is a three-stage process. It involves a printing step, a de-binding step, and a sintering step. Here's how the printing step works:

- The raw material comes in a filament or rod form that typically consists of metal particles bound together by polymer and/or wax.

- This rod or filament is extruded through a heated nozzle and deposited layer-by-layer building a part based on the CAD model.
- At the same time, support structures are built if necessary. The interface between the support and the part is printed with a ceramic support material, which is easy to manually remove later.

Fig. 9.1: Metal Parts Manufactured by Fused Deposition Modeling [192]

When the print is complete, the resulting "green" part needs to be post-processed to become metal using similar steps as in Binder Jetting. The "green" is first washed in a solution for several hours to remove most of the binder. Then it is sintered in a furnace to bond the metal particles together and form a fully-metal part. During sintering the part dimensions are reduced by about 20%. The parts are printed larger to compensate for this. Like in Binder Jetting, this shrinkage is not homogenous. This means that some trial and error is needed to produce accurate results for a particular design. Figure 9.1 shows a metal component manufactured by fused deposition modeling.

9.4. Key Considerations

Metal Additive Manufacturing is a booming industrial sector and an active field of research. Even through it can't fully replace conventional manufacturing techniques, laser or electron-beam-based metal AM processes have known significant developments in the past years. In some cases, metal additive manufacturing offers a relevant technical-economic solution, for instance for components with an asymmetry between design complexity and production volume.

9.4.1. Existing Designs May Not be Directly Transferable

Due to its high cost, it is rarely economically feasible to manufacture parts that have been designed for a traditional process using metal 3D printing. Actually, it is often technically impossible to reproduce these geometries. For example, sections thicker than 10 mm are prone to warping or other manufacturing defects and should be avoided.

9.4.2. Geometric Complexity is as Asset

Design complexity is often considered harmful, as it is connected to an increased cost. In metal 3D printing, this is not the case. On the contrary, finding a way to maximize the added value that geometric complexity brings to a system is key to take full advantage of the benefits of metal 3D printing.

9.4.3. Start by Defining the Basic Requirements

When you begin to redesign a part or assembly for metal 3D printing, it is usually a good idea to start with a blank canvas. This way you can avoid being restricted by preconceived designs. Clearly defining the design requirements (loads, boundary condition, part weight etc.) is key here. We will see in the next section that modern CAD software use these as input to create optimized structures with organic forms.

9.4.4. Always Use Minimum Support Structures

It is a good practice to have a clear vision of how the part will be orientated in the machine. Print orientation is important as it defines the position and need of support structures. The aim of the designer should be to create parts with self-supporting features, minimizing the need for support and ensuring build success.

9.4.5. Post-processing is Always Required

Independent of the process, post-processing is always required in metal 3D printing. This can be obligatory (such as support removal in DMLS/SLM or sintering in Binder Jetting and Metal extrusion) or optional (such as CNC machining step to achieve tighter tolerances or a heat treatment to improve material properties). So it is essential to keep the post-processing requirements and available option in your mind while designing a part for metal 3D printing [193].

9.5. Creating Metal Parts by Fused Deposition Modeling and Sintering

Conventional metal 3D printings are generally high energy consuming and expensive. As an alternative, Fused Deposition Modeling and Sintering (FDMS) was proposed for fast making metal parts at low energy consumption and low cost. Metal/polymer composite filament is printed by the printer based on Fused Deposition Modeling (FDM), and then debinding and sintering are conducted on the printed parts to form compact metal parts. Fused Deposition Modeling and Sintering (FDMS) is a novel AM method based on FDM in which filaments are metal/polymer composites with a high loading of metal powders, and the printer is further optimized on the basis of the conventional FDM printer. Figure 9.2 shows a schematic illustration of the FDMS process.

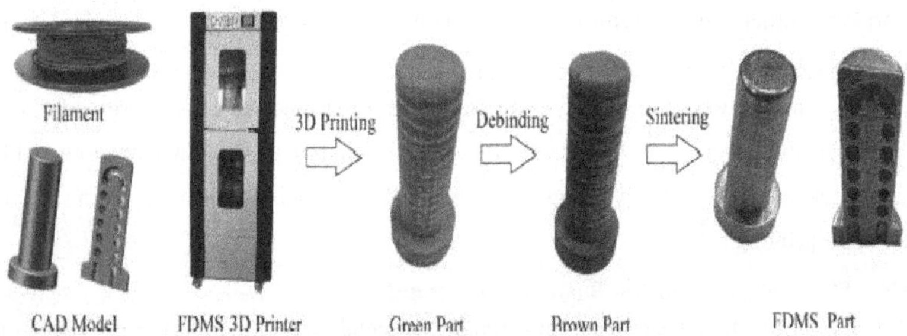

Fig. 9.2: Schematic Illustration of FDMS [194]

For the sake of simplicity, we define the printed part as 'Green Part', the debinded part as 'Brown Part', and the sintered metal part as 'FDMS Part', respectively. Firstly, Green Parts are printed from metal/polymer composite filament by FDM, during which polymer is melted as the binder but the metal particles remain solid. Later, Brown Parts was obtained by subjecting the Green Parts to a debinding process to remove most of the polymer binder. The rest polymer binder in the Brown Parts can avoid the spreading of the metal particles and thus preserve the shape of the parts. Finally, the Brown Part is sintered to fuse the metal particles together to form dense FDMS Parts.

Bin et al [194] in their experimental work fabricated metal/polymer composites filament, consisting of a polymer matrix and dispersed 88 wt% 316L stainless steel particles with size in 30–50 lm. The polymer matrix is composed of polyformaldehyde (POM) and additives such as polypropylene (PP), dioctyl phthalate (DOP), dibutyl phthalate (DBP), and zinc oxide (ZnO) to increase the fluidity, plasticity, and thermo-stability of the composite. The melt flow rate of the composite filament at 230°C with a load of 2.16 kg is 23.56 g per 10 min.

9.5.1. *Experimental Methods*

First, using a CoLiDo metal 3D printer from Print-Rite Unicorn Image Products Company Limited of Zhuhai, Green Parts were fabricated by the optimized process parameters with a layer thickness of 0.2 mm, a printing temperature of 230 ^{0}C, and a filling pattern of perimeter shells and rectilinear internal filling. Then, catalyst debinding to achieve the Brown Parts was conducted to remove the POM binder in the Green Parts at 120 0C for 8 h under nitrogen whose rate is 1 L/h. Finally, the Brown Parts were subjected to a sintering process by first removing the remaining binder apart from POM at 600 0C for 2 h and then sintering at a temperature of 1360 ^{0}C for 2 h under the protection of argon gas.

9.5.2. *Microstructure Analysis*

Figure 9.3a shows the scanning electronic microscopic images of the cross-section of the filament used in FDMS. The binder in the filament forms a continuous phase and the 316L particles with different sizes are uniformly dispersed in the binder matrix, as indicated by the arrows and other microspheres in Fig. 9.3a. Since the binder plays a principle role in the printing process, the requirement on the morphology of the dispersed metal particles, including the sphericity, fluidity and size distribution, is not as rigorous as that for SLM or other metal 3D printing methods. Thus, the metal particles used in this filament for FDMS are much easier to be obtained. Due to the same reason, any metal with a melting point above 300^{0}C can be applied in FDMS. Fig. 9.3b shows the cross section of a manufactured Green Part. Since the flat filiform composite extruded from the nozzle cannot seamlessly fuse with the deposited bottom layer, slit-shaped printing gaps were found, which can be attributed to the fast solidification of the extruded melt. The filling pattern of rectilinear is 45^0 inclined to the horizontal, so the printing gaps in the internal zone A and zone B in Fig. 9.3b is longer and wider than that in C perimeter shell zone. Furthermore, the slicing software can set a few bottom layers as solid layers leading to the printing gaps in zone D of Fig. 9.3b much smaller and leaner.

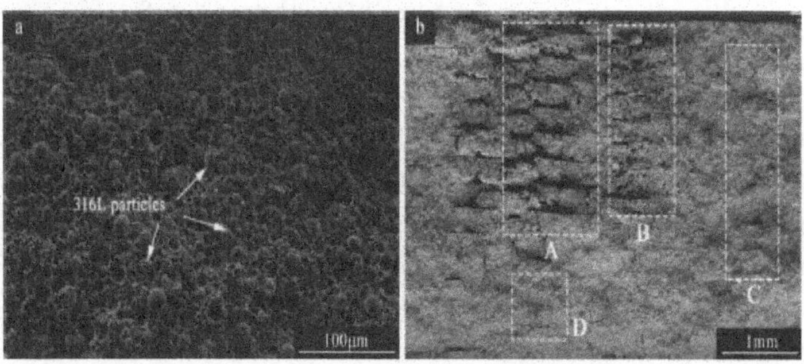

Fig. 9.3: SEM Images of the Cross-sections of a) Filament and b) Green Part [194]

Figure 9.4 shows the microstructure of the FDMS 316L Part after the debinding and sintering processes. Pores at micron level formed and uniformly distributed throughout the cross-section, and most of them are preferentially spherical. The tiny printing gaps in the shell (zone C in Fig. 9.3b) and bottom layers (zone D in Fig. 9.3b) of the Green Part has been repaired during the sintering process. Although printing gaps can still be observed in the areas with relatively big gaps, as shown in zone A and zone B in Fig. 3b, they shrunk after the sintering process and are much smaller than those in the Green Part (Fig. 9.3b). More specifically, the bigger printing gaps in zone A turned to closed narrow cracks, the smaller ones in zone B turned to irregular pores with sizes at several tens microns. The inherent flaws, especially the printing gaps, become potential cracking sources, which will result in the lower strength of the FDMS Parts. A further parametric research to optimize the printing and sintering process should be carried out in the future to reduce the pores and internal defects. After sintering, a full austenite phase with twin crystals was achieved in the FDMS 316 Part as shown in Fig. 9.4b.

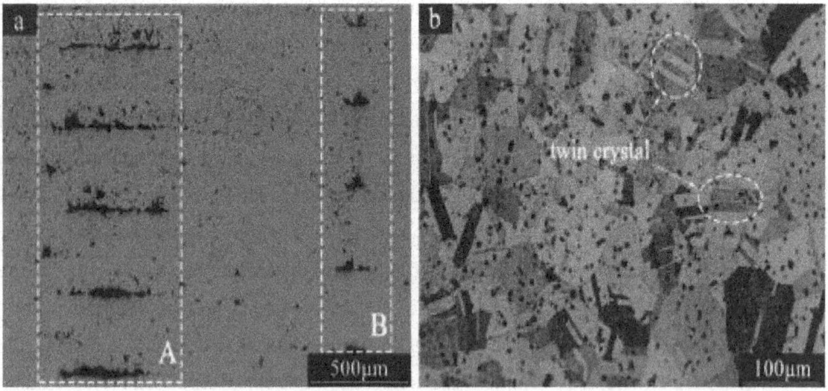

Fig. 9.4: Optical Microscopy of Microstructure in FDMS 316L Part [194]

9.5.3. *Hardness and Tensile Strength*

The mechanical properties were obtained based on the averaged value of five sintered FDMS 316L specimens. The micro Vickers hardness of the FDMS 316L Parts is 145.2 ± 6.7 HV, which is slightly lower than that of AISI type 316L parts (155HV), but significantly lower than that of SLM 316L parts (232HV). Although the formed twin crystal (Fig. 9.4b) during the sintering process can improve the hardness, the relatively high porosity dramatically decreases the hardness since the FDMS Parts with the pores or the printing gaps can be easily pressed into by the indenter of the hardness tester.

The tensile properties were measured for the FDMS Parts and were compared with those of metal parts manufactured using other techniques. It can be found that the FDMS 316L Parts possesses lower mechanical properties in terms of the yield strength, ultimate tensile strength, and elongation at break, especially for the ultimate tensile strength. Considering the relatively lower mechanical properties of the final products from FDMS, this forming process is unsuitable to manufacture structural parts, but suitable to fabricate functional products with relatively low requirements on mechanical performance.

9.6. Fabrication of Iron - Nylon Composite Filament

Masood et al [195] in their experimental work fabricated iron Nylon composite filament. The filaments used in FDM process need to be of specific size, strength and properties. A single-screw extruder machine was used to produce such filaments. The single screw extruder is one of the most important types of extruders used in the polymer processing industry. Its key advantages include its relatively lower cost, simple design, ruggedness, reliability, and acceptable performance [196].

A Brabender Plasti-Corder single screw extruder was selected to fabricate the filaments from the composite material. The composite mixture powder is fed into the feed hopper. The mixture powder material flows by gravity from the feed hopper down into the extruder barrel. As the material flows, it fills the annular space between the extruder screw and barrel. Since the barrel is stationary and the screw is rotating, the frictional forces will act on the material, both on the barrel as well as on the screw surface. As the material moves forward, it will heat up as result of frictional heat generated and conducted from the barrel heaters.

When the temperature of the nylon material exceeds the melting point, a thin film of composite material, which is a compounded nylon and iron, will form at the barrel surface. This is where the plasticising action starts. As the composite material moves forward, the amount of solid material at each location will reduce as a result of melting. When the entire solid composite has disappeared, and the end of the plasticising zone has been reached then the melt-conveying zone starts. In the melt-conveying zone, the composite material is simply pumped to the die. Since the die exerts a resistance to flow, a pressure is required to force the material through the die. This is referred to as the die head pressure. The die head pressure is determined by the shape of the die, temperature of the composite material melt, the flow rate through the die, and the rheological properties of the composite melt.

The extruding filament exiting the die is then carried away by a conveyor belt, whose speed matches the output flow rate. During extrusion, a phenomenon, sometimes referred to as die swell, occurs. Actually, it is not the die but the polymer that swells. The elastic behaviour of the polymer melt is largely responsible for the swelling of the extrudate upon leaving the die. This is primarily due to the elastic recovery of the deformation of the metal-polymer composite in the die. A die with a short land length will cause a large amount of swelling, while a die with long land length will reduce the amount of swelling. Therefore, the land length of die was increased from 5 to 10 mm, which leads to decrease the extrudate swelling. Because of swelling, the diameter of die used was taken to be smaller than the required diameter of the filament. The die diameter was 1.65 mm and the resulting diameter of filament formed was in the range of 1.78–1.85 mm due to the extrudate swelling. Therefore, the geometry of exit flow channel is generally different from the required product geometry. Thus using the proper geometry of the die and appropriate volume of flow parameters, flexible winding filaments of desired diameter and tolerance were produced. Figure. 9.5 shows the final filament produced by this process, ready for use in the FDM machine.

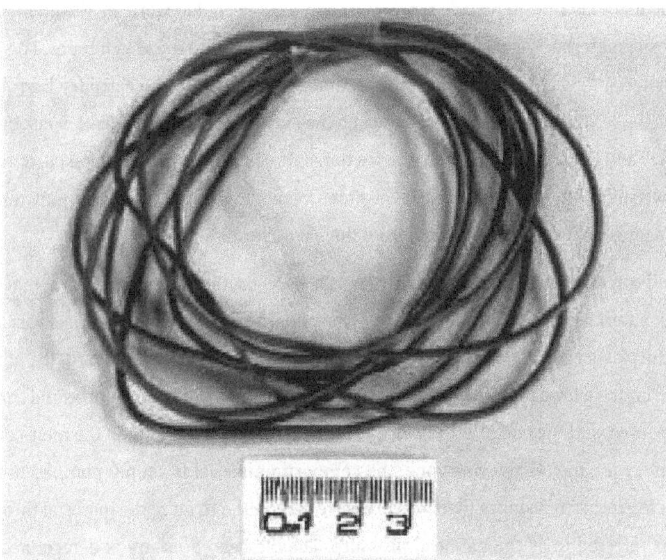

Fig. 9.5: Filament Produced by Iron – Nylon Composite [195]

9.6.1. *Tensile Properties*

The quality of feed stock filaments is critical to the proper fabrication of FDM components. Successful processing of composites in FDM starts with fabrication of filaments with a uniform particle dispersion, possessing certain mechanical properties, rheology and adhesion behaviour.

When the volume fraction of metal particles and the size of the particles in the polymer matrix are varied, the tensile stress, tensile modulus, and elongation of the metal composites are affected. From microscopic studies, it has been noticed that there is no strong bond adhesion between the filler particles and the polymer matrix. This may be due to cracks or voids forming around the filler particles in composite. The specimens of the filament were tensile-tested in an Intron testing machine, corresponding series IX Automated Materials Testing System 1.26.

The testing was carried out using a standard string test with a string clamp. All the tests were performed with a crosshead movement of 20 mm/min. Four specimens were tested in each case. Table 9.1 shows the results of tensile modulus, tensile stress at peak and tensile strain at break for each of three samples of iron/nylon composite filaments.

Table 9.1: Tensile Modulus, Tensile Stress at Peak and Tensile Strain at Break [195]

Specimen no	**Constituents by volume**	**Tensile modulus E (MPa)**	**Tensile stress at peak σ (MPa)**	**Tensile strain at break € (%)**
1	30%Fe (coarse) 70% nylon	54.52	3.87	16.82
2	40%Fe (coarse) 60% nylon	42.23	2.76	12.94
3	40%Fe (fine) 60% nylon	37.18	2.44	74.61

The values shown are the average values obtained from four tests in each case. Figure 9.6 shows the stress–strain curves of iron/nylon composites with different volume fraction and particles size of iron. In this case, the stress is defined as the tensile load divided by the cross-section of the filament and strain is defined as the fractional change in the length of the filament.

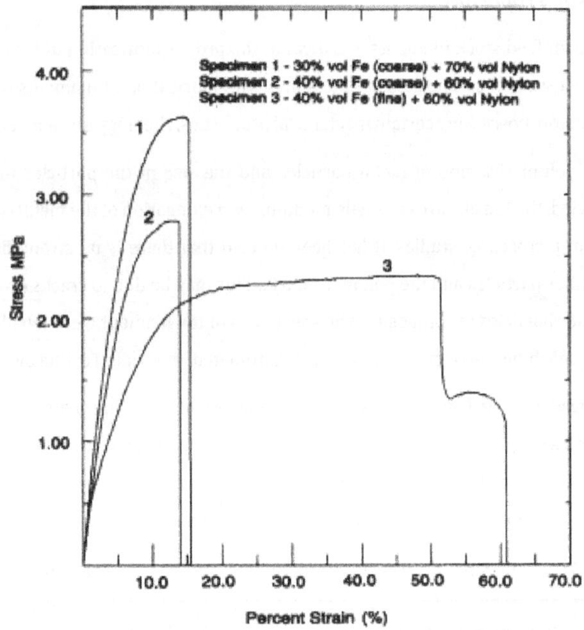

Fig. 9.6: Comparison of Stress Strain Curves for Iron Nylon Composites [195]

In figure 9.6, the initial linear portion of the stress–strain curves close to the origin represents the elastic elongation of composites. During initial stress, the iron particles were bonded by compression by the nylon matrix due to large difference in thermal expansion coefficients. Once the applied stress overcomes the residual compressive stress, the nylon matrix surface normal to the applied tensile force separates from the iron particle surface to form a pseudovoid. The stress, at which the pseudovoids are formed, depends on both the degree of the interfacial bond and the magnitude of the compressive stress.

Once they are formed, the material becomes a three-phase composite, consisting of polymer phase, bonded particles, and pseudovoids. Since the pseudovoids no longer support the applied tensile force, they cause the modulus of the composite to decrease. At approximately 60% of tensile stress for specimens 1 and 2, the slope begins to decrease quickly until fracture. The slope of specimen 3 is flatter than those of specimens 1 and 2 at approximately 80% of tensile stress. For specimens 1 and 2, Fig. 9.6 also shows the variation of the modulus of elasticity (as the slope of the linear portion) and the tensile stress for composites of different volume fractions of iron content.

The experimental results show that higher volume fraction of composite represents lower modulus and tensile stress at volume fraction greater than 0.30. When the volume fraction of iron particle is increased, the decrease in fracture energy occurs. This is explained as being caused by an ineffective interaction when the particles become closely spaced. The composite appeared as a continuum and the crack front did not interact with the discrete particles. Another possible reason is that the crack precursors are increased with increasing fraction of particles. Each particle of the dispersed phase can be considered as a crack precursor during stressing. The incorporation of voids, which is unavoidable in fabrication of composites, and cracks formed during stressing, lower the modulus and the stress level. Also the bond between the iron particles is not as strong as that between the nylon matrix and iron particles. This may be due to voids formed with the application of tensile load. The tensile load cannot properly transfer stress thus lowering the values of mechanical properties of composite material. Somes examples of parts manufactured by FDM process is shown in figure 9.7.

Fig. 9.7: Examples of Parts and Inserts Produced on the FDM System [195]

9.7. Metal-Polymer Composite Filament

Seyeon et al fabricated thermoplastic ABS composite with copper and iron powder. The size of copper and iron particles was less than 24 μm and 43 μm respectively. The contents of metal powder were varied from 10 wt.% to 50 wt.% in order to observe the effects of metal particles on thermomechanical properties of the final products. The metal/ABS mixture was formed into pellets, and the pellets were extruded by a filament extruder (Filastruder) to make a metal/polymer composite filament for a FDM 3D printer. The final diameter of the filaments was 1.75 mm. Metal filaments were printed by a NP-Mendel (Opencreators) which is a type of FDM 3D printer. Dog bone-shaped structures were manufactured to analyze their tensile properties [197].

9.7.1. *Effects of Viscosity on Tensile Properties*

The melting temperature (Tm) of ABS was measured by a differential scanning calorimetric analysis using a DSC8000 (Perkin Elmer). According to Figure 9.8, the melting temperature of ABS is 2170C. The glass transition temperature is estimated to be 173.6^{0}C because the Tg is commonly 80% of the melting temperature.

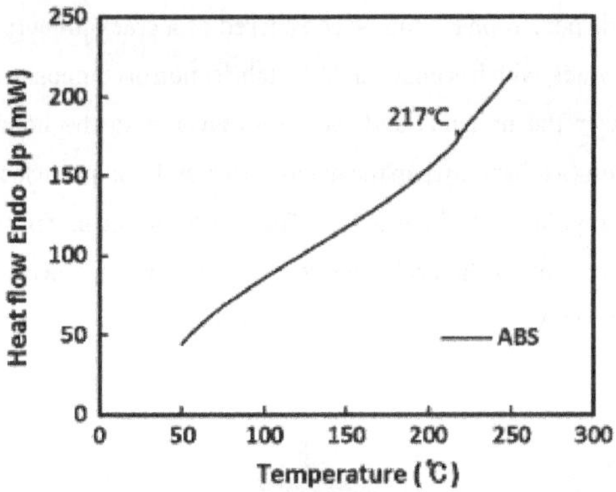

Fig. 9.8: DSC Analysis Result of ABS [197]

Table 9.2 shows how the volumetric flow rate during the extrusion of ABS increases as the temperature is increased. The volumetric flow rate, Q, was increased from 1.82 x 10-9 m^3/s to 3.47 x 10-9 m^3/s over the temperature range shown in the Table. We can conclude from this data that the viscosity of ABS decreases as the temperature increases.

Table 9.2: The Volumetric Flow Rate and Viscosity of ABS at Various Temperatures [197]

Temperature (^{0}C)	Volumetric flow rate, Q (m^3/s)	Viscosity, μ (cP)
190	1.82 x 10 $^{-9}$	3,980,000
200	2.38 x 10 $^{-9}$	3,.050,000
210	2.94 x 10 $^{-9}$	2,460,000
220	3.47 x 10 $^{-9}$	2,090,000

The viscosity of ABS at various temperatures was calculated by the use of the volumetric flow rate and the Hagen–Poiseuille equation. The values of the parameters such as radius of nozzle (r), length of nozzle (L), and pressure difference (DP) were 0.875 mm, 21 mm, and 6.74 x 10^8 kgf/cm^2, respectively. As depicted in Figure 9.9 and Table 9.2, the viscosity of ABS was decreased from 3,980,000 to 1,630,000 cP when the temperature was changed from 190°C to 220°C.

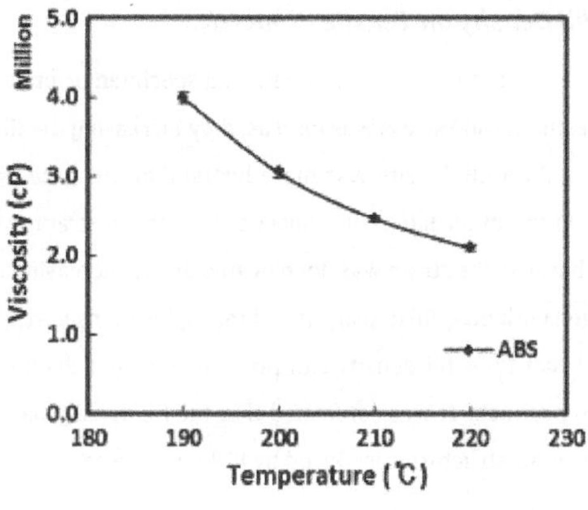

Fig. 9.9: Viscosity (μ) of ABS at different Temperatures [197]

Figure 9.10 shows the measured stress–strain curve for the final 3D products manufactured by FDM at different temperatures. The stress required to deform the structures increased with increase in temperatures. Likewise, the tensile strain increased as the temperature was raised. It is theorized that the adhesion between the layers at lower temperatures is weaker than at higher temperatures, resulting in decreasing tensile strength and strain of the 3D printed structures. This result can be explained in terms of the viscosity of ABS. As mentioned above, the viscosity of ABS is reduced as temperatures increase; hence, the tensile strength of the 3D structures printed with FDM is decreased by reducing the viscosity of ABS as it is extruded.

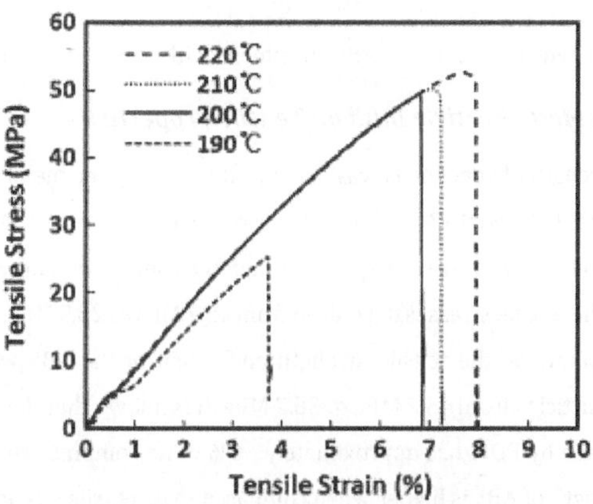

Fig. 9.10: Tensile Properties of the 3D Structures Manufactured by ABS at different Temperatures [197]

9.7.2. *Effects of Fill Density on Tensile Properties*

Figure 9.11 allows us to see the tensile strength of a specimen printed using FDM with ABS filaments. In this case, the tensile stress was increased by increasing the fill density. That is, the specimen printed with 20% fill density was more brittle than the other specimens printed at higher fill densities. On the other hand, the aspect of the tensile strain was different from the case of tensile stress because the strain was not elevated due to increasing the fill density. When a specimen was printed with 60% fill density, it had the highest tensile strain, so it was the most ductile. It is believed that 60% fill density can provide sufficient ductility for manufacturing functional 3D printed products. It was confirmed that the fill density has an important role in the tensile properties of 3D structures produced by FDM processes.

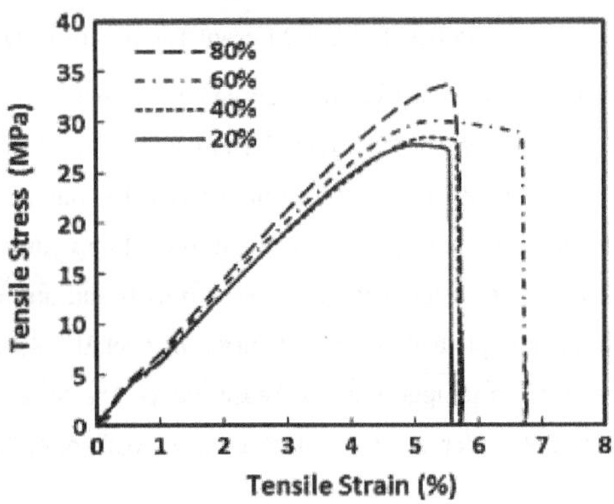

Fig. 9.11: Tensile Strength of ABS Specimens with Increasing fill Density[197]

9.7.3. *Effect of Metal Particle Infill on Tensile Properties*

The tensile strength of specimens was decreased by raising the metal contents. This can be observed in figures. 9.12 and 9.13 and Table 9.3. The tensile stress of ABS without metal particles was 45.7 MPa, and its tensile strain was 8.3%. When the copper content was increased from 10 wt% to 30 wt%, the tensile stresses decreased from 45.7 MPa to 26.5 MPa, and the tensile strain also decreased. Likewise, the tensile strength and strain of the ABS was reduced with the addition of iron particles from 45.7 MPa to 36.2 MPa. It is known that the tensile strength of an ABS specimen made by FDM has approximately 66% of its compressive strength. That is, the compressive strength of ABS is higher when compared to its tensile strength. For this reason, it is believed that metal/polymer composite filaments could be applied to make structures requiring resistance to compressive force rather than a tensile force.

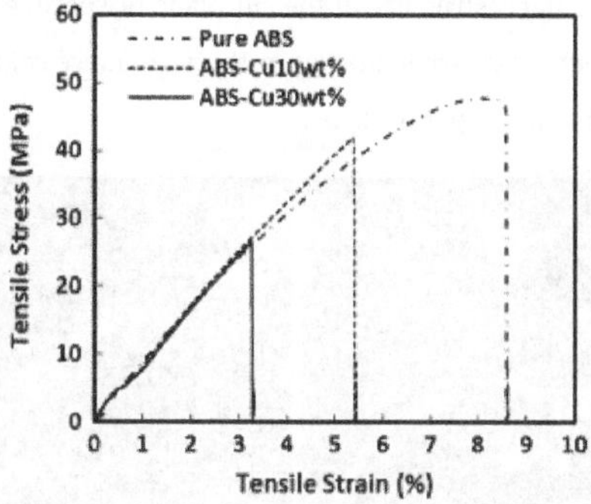

Fig. 9.12: Stress–strain Curves for ABS with Copper Particles [197]

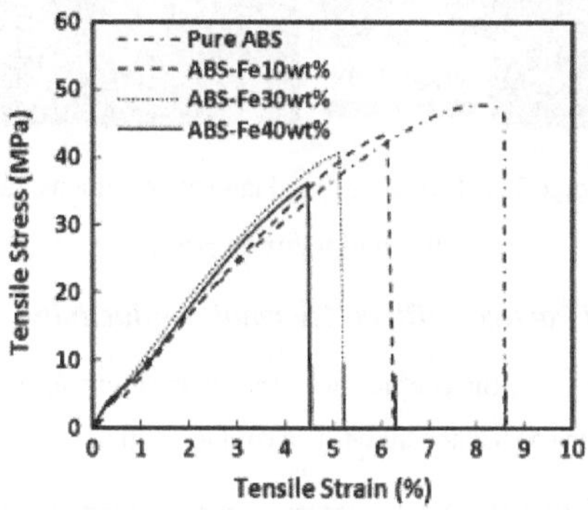

Fig. 9.13: Stress–strain Curves for ABS with Iron Particles [197]

This results from lower contents of ABS with the use of 30 wt% of copper particles. Many voids cause lower binding force between layers. It is also believed that copper particles in the ABS matrix disturb the bonding force between the layers. This is why the tensile strength of ABS-Cu 10wt% is larger than that of ABS-Cu 30wt%.

Table 9.3: Tensile Properties of the Specimen [197]

Composition (wt%)	**Tensile modulus, E (MPa)**	**Tensile stress, σ (MPa)**	**Tensile strain, € (%)**
Pure ABS	880.4	45.7	8.3
ABS - Cu 10	930.2	42.0	5.4
ABS - Cu 30	915.9	26.5	3.3
ABS - Fe 10	906.1	43.4	6.1
ABS - Fe 30	978.5	40.6	5.1
ABS – Fe 40	958.2	36.2	4.5

Figure 9.14 shows the microstructure of the cut plane of specimens which is made by the ABS-Cu composite filament. As shown in the figure, there are more voids with 30 wt% of copper particles than with 10 wt.% of copper particles.

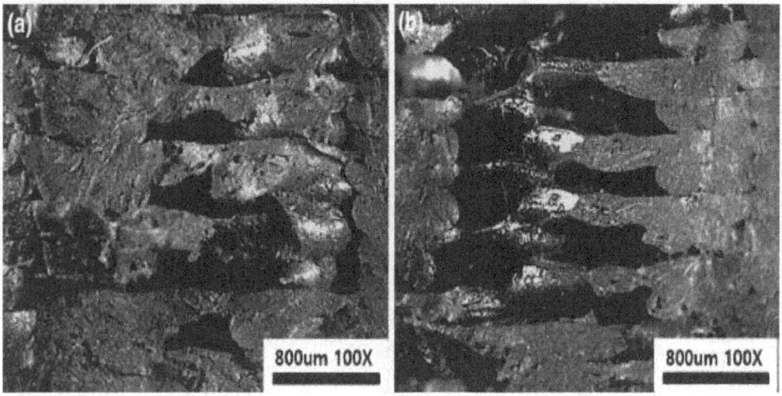

Fig. 9.14: Microscopic Structure of Cutting Plane of Specimens: (a) ABS-Cu10wt% (b) ABS-Cu30wt% [197]

9.7.4. Effect of Metal Particle Infill on Thermal Conductivity

As depicted in figure. 9.15, the coefficient of the thermal expansion of the specimens was decreased to 29.5% of pure ABS when 50 wt.% of copper particles was added into the ABS.

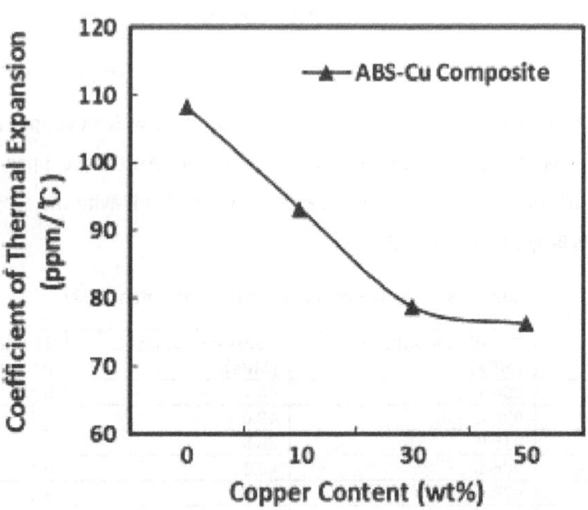

Fig. 9.15: Coefficient of Thermal Expansion by Changing the Copper Content [197]

Specifically, the CTE was reduced from 108.2 ppm/°C to 76.2 ppm/°C with increase in content of copper particles. This shows that the distortion problem of the final 3D product can be prevented by using the ABS-Cu composite filaments. The thermal diffusivity, heat capacity, and density of the materials were used to find the thermal conductivity of the final 3D structures. As shown in figure 9.16.

The thermal conductivity was increased from 0.646 W/mK to 0.912 W/mK when the copper content in ABS was increased from 0 wt% to 50 wt%. This improvement of the thermal conductivity with use of copper particles illustrates that the metal particles play an important role in the FDM industry as a function of printable materials, e.g. the more metal particles, the better the thermal conductivity of the FDM filaments.

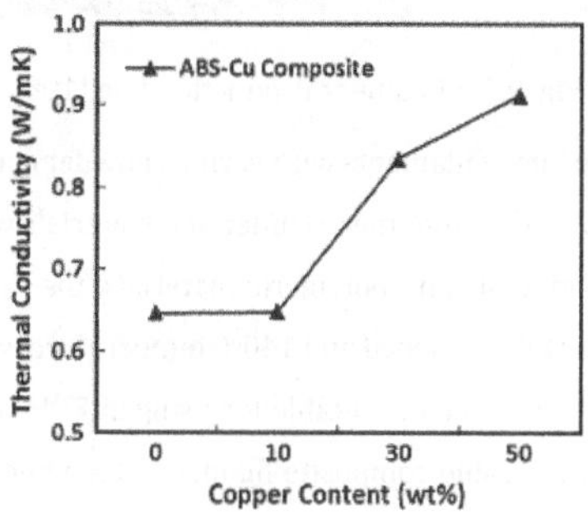

Fig. 9.16: Thermal Conductivity of ABS-Cu Filaments as a Function of Copper Content [197]

Consequently, it is believed that the ABS-Cu filament would provide the opportunity to make large-scale 3D electromagnetic structures, such as antennas, with no distortion of the final product.

9.8. Polymer Metal Composite (PE/Cu)

Nabipour et al in their experimental work manufactured polymer/metal composite (PE/Cu) parts via FDM method to produce high-quality filaments. In their study, granular PE of BL3-class produced in Jam Co. (Jam, Iran), and copper powder made by Purian Institute of Chemistry (Tehran, Iran) in size smaller than 20 μm was used as the polymeric and metallic materials, respectively. Additives such as paraffin wax and stearic acid were also used to lubricate and increase the quality and flexibility of the produced filaments. The production line shown in figure 9.17 consist of a corotating twin extruder with 40 mm in diameter for mixing, an orifice die to shape the filament, cooling water bath, and the drawing equipment to fix the final size of the filament.

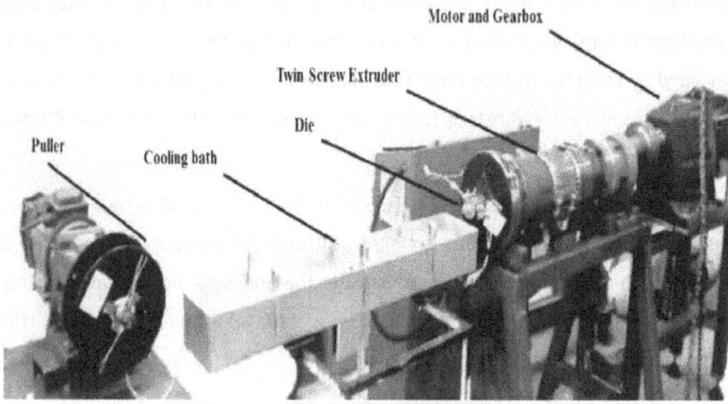

Fig. 9.17: Filament Production line [198]

The installed production line of filaments with various powder contents; weight percentages of 25, 50, and 75%. Prior to feeding into the extruder, the materials were dry mixed. In order to attain a quality mixture of metallic and polymeric materials, the raw materials and additives were mixed for 40 minutes at 80 rpm speed and 140°C temperature with a handmade laboratory mixer. The typical diameter of filaments suitable for using in FDM 3D printing process is 1.75 mm, but in this study, the preferable composite filaments diameter was found in the range of 1.8–1.9 mm. Figure 9.18 depicts a schematic view of filament die and Figure 9.19 shows the produced filaments, also optimum production parameters are given in table 9.4.

Table 9.4: Optimum Parameter for Filament Production [198]

Filament diameter (mm)	**Puller speed (rpm)**	**Extruder speed (rpm)**	**Nozzle diameter (mm)**	**Nozzle temperature (°C)**
1.85±0.5	11.5	8	4	225

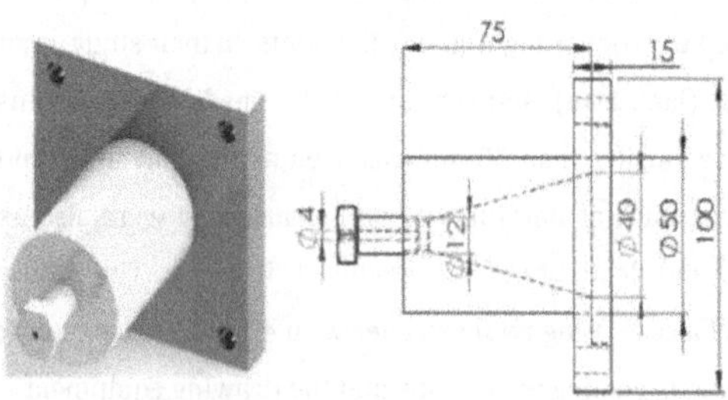

Fig. 9.18: Schematic Illustration of Filament Production Dies [198]

The first obstacle to feed the manufactured filaments of PE was due to its high flexibility that was tackled by slight increase in diameter (in the range of 1.8–1.9 mm) and a decrease of distance between the feeding gear and the liquefier; these two parameters (diameter and length of a column) could increase the critical stress for buckling. Generally, in the FDM process, "the bed" is made of glass or metallic materials coated by fireproof adhesive glue. A critical problem for printing PE-based specimens in the FDM process is the lack of adhesion to this substrate. Preliminary efforts were made to print PE on the substrates at different temperatures that resulted in faulty adhesion. As an alternative, a thin sheet of PP was used to address this issue. PP is a semicrystalline thermoplastic, with the most similar properties to PE, but with a higher melting point, that was as a coating on the bed to supposedly accommodate the adhesion of the PE molten material to the substrate. However, as the number of deposited layers on each other increased, due to intrinsic warpage and shrinkage of PE, the part, along with the PP sheet, was detached from the bed surface.

Overall, the attachment of the part to the thin sheet of PP was not completely intact. This is clearly shown in figure 9.19.

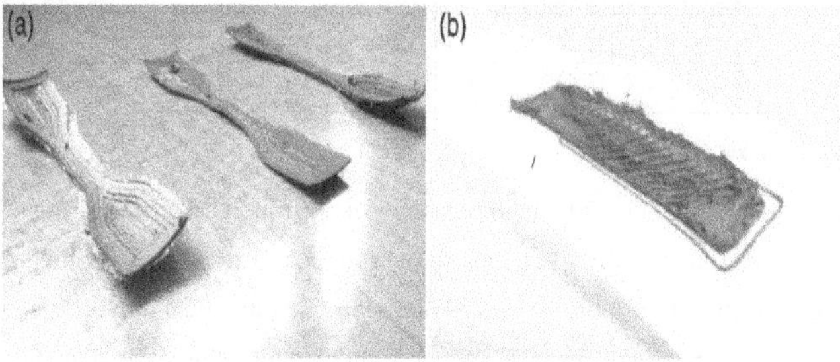

Fig. 9.19: (a, b) Distortion of PE-based Specimens [198]

Generally, the interlayer adhesion in the FDM process for semicrystalline materials is much stronger than the amorphous materials. Amorphous materials begin to soften in their glass transition temperature, whereas semicrystalline materials experience melting rather than softening, which leads to a relatively stronger adhesion. As a remedy to ease the part detachment, some concentrated efforts were made to optimize the quality of attachment of the printed part to the substrate, via variation of printing parameters. In another word, one could regulate the amount of attachment to an optimum so that it guarantees a better attachment of the part to the substrate during deposition as well as ease of detachment by the common tool. First layer printing speed and nozzle diameter are two crucial parameters which have vital roles in achieving an optimal adhesion.

Secondly, nozzle diameter is another effective parameter, which not only can provide the optimal strength of adhesion between the part and the bed but also have an influential impact on blocking during the process. In fact, with an increase in nozzle diameter, the thickness of the printed rasters increases and allows for stronger adhesion between the part and the bed. On the other hand, when the adhesion is stronger, detaching of the final part from the bed becomes more difficult. Moreover, by adding metal particles to the neat PE, it is possible for the nozzles with lower diameters to block and clog during the process.

Consequently, the optimum value for the nozzle diameter was experimentally calculated as 1 mm in this section. A 1 mm nozzle can provide good adhesion between the part and the bed both to ensure that the part is not detached during the process and the ability to easily separate the final part at the end without any clogging during the process. Figure 5 shows a desirable PE specimen printed by a 3D FDM printer and the optimal printing conditions of the specimens are given in table 9.5.

Table 9.5: Optimum Condition for Printing [198]

No	Process parameters	Optimum condition
1	Bed material	Rough (hatched) PP plate (THK= 6mm)
2	Bed temperature (°C)	25
3	Nozzle temperature (°C)	250
4	Nozzle diameter (mm)	1
5	First layer print speed (mm/s)	20
6	Overall print speed (mm/s)	30

9.8.1. Flexural Strength Test

Specimens with 3.2 × 13.5 × 80 mm dimensions were printed according to the ASTM-D790 standard for the flexural test. The specimens were produced with copper weight percentages of 0, 25, 50, and 75% by FDM 3D printer. As can be seen in figures 9.20 and 9.21, by increasing copper particles, the maximum flexural strength and modulus increase, and the maximum flexural strain decreases.

Based on previous studies, the tensile strength, max elongation, and modulus decreased with metallic particles content owing to the separation of filler from the polymer matrix during the test. On the contrary, it is interesting to notice that the flexural strength and modulus increased with copper content. It is clear that in the three-point bending test, the bottom surface undergoes tension whereas the loaded surface undergoes compression on the opposite side. This condition is shown in Figure 9.22. It is also clear that in tension section any defect or flaw expand in the part body, while in compression section the defects cannot lead to a weak structure but can be plugged because in the flexural tests there is low deformation. So it is conceivable that in these composite parts, the defects could not weaken the structure owing to lack of enough time to take effect.

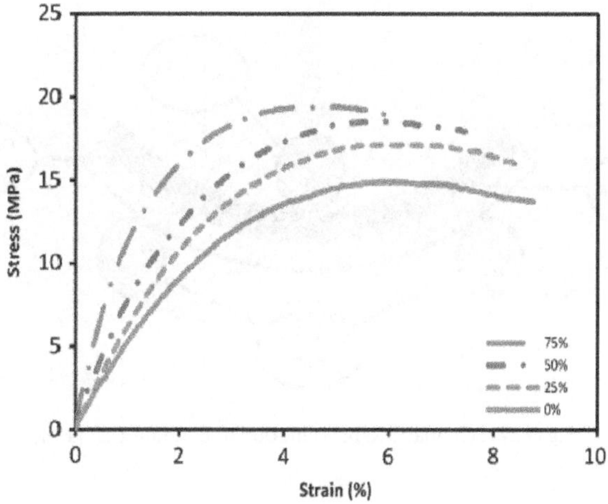

Fig. 9.20: Flexural Strength Test [198]

Therefore, the part body indicates rigidity behavior in the bending test. Indeed, metallic particles are rigid and incompressible, so when added to the PE matrix, reduce the amount of space for compressive deformation.

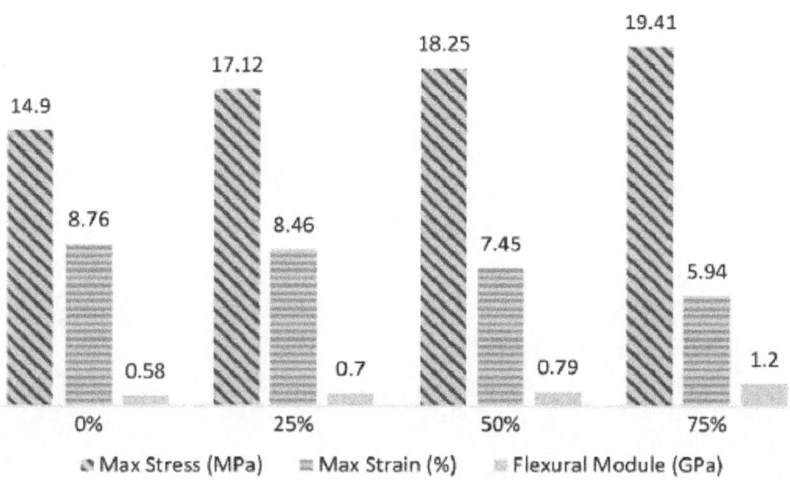

Fig. 9.21: Flexural Strength Test Results [198]

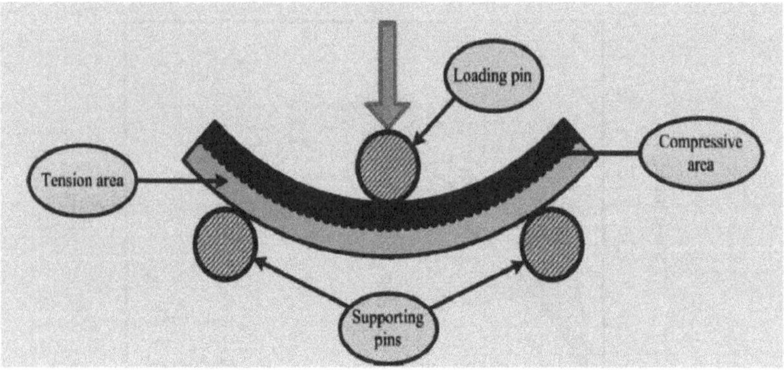

Fig. 9.22: Schematic Representation of Bending Test [198]

CHAPTER 10

10. Future Research Scope and Conclusion

10.1. Future Research Scope

In FDM different process parameters have effects on the part quality of FDM. Different process parameters have effects on the part quality of FDM. Recent years, research has been targeting into identifying the optimal process parameters to improve surface finish, aesthetics, mechanical properties, model material consumption and build time. However, there are still no perfect optimal conditions for all types of parts and materials. For most parts, there is always a need to adjust parameters to balance a tradeoff between production time, surface finish, and dimensional accuracy. The properties of the FDM fabricated parts can be controlled by the selected build styles and other FDM parameters. Recent years, research has been targeting into identifying the optimal process parameters to improve surface finish, aesthetics, mechanical properties, model material consumption and build time. However, there are still no perfect optimal conditions for all types of parts and materials. For most parts, there is always a need to adjust parameters to balance a tradeoff between production time, surface finish, and dimensional accuracy. The properties of the FDM fabricated parts can be controlled by the selected build styles and other FDM parameters. Thus, new mathematical modeling approaches and optimization techniques need to be developed. After reviewing past research works relevant to the areas conducted by the researchers, where main focus was on the optimization of process variables for FDM, the research work leaves a wide scope of improvements and hence more future researches can be performed to explore many aspects of FDM. The following observations are made for future research.

The literature review indicates that much research work has been attempted to improve the mechanical properties and part quality for FDM fabricated ABS parts by optimizing one or several important process parameters. From previous studies, it has been shown that the quality of FDM built part is highly affected by various process variables. Hence, the identification of the critical process parameters and determination of optimum process parameters can lead to the quality improvement of FDM fabricated part. However, the relationships between the process parameters and the part quality and mechanical properties have not been studied enough especially for various types of materials used by the FDM process. It remains a matter of concern that there are no absolute rules and guidelines designed to assist in their optimization and evaluation method.

The FDM liquefier head is designed to be capable of dealing with different kinds of materials such as ABS, polyphenylsulfone (PPSF), PC-ABS, PC, PC-ISO, nylon-12, elastomer and wax. Amongst them, PC, PC-ISO, PC-ABS and PPSF are widely used for FDM process and their applications are growing rapidly in the global market of manufacturing. Many studies have investigated the effects of FDM process parameters on ABS built part. However, in case of other FDM materials, very little work has been done both in terms of material characterization and FDM process optimization. Therefore, considerable work remains to be done in DOEs for part fabrication and process optimization involving other FDM polymers such as PC, PPSF, PC-ABS, PC-ISO, elastomer and nylon-12.

In terms of material properties, most of the studies are mainly focused on optimizing the process parameters for mechanical properties of ABS parts. However, there have been no published research articles relating to the optimization of FDM process variables for thermal, chemical and dynamic mechanical properties of FDM fabricated parts in other material forms. Therefore, much research work is needed in this area in the future research.

Effects of process parameters on surface roughness, tensile strength, compressive strength, flexural, impact strength and dimensional accuracy have been studied. But the study needs to be extended to other types of quality characteristics such as hardness, production time, creep, vibration, product and process cost, porosity and stress strain behavior at high-strain-rate loading conditions. Future work should also focus on the application of new statistical designs, modeling and optimization techniques to find out the optimal combinations for a certain set of process parameters and test the functionality of FDM processed part in an efficient manner.

Environmental factors such as temperature and relative humidity are also considered to be the sources of error affecting part accuracy. FDM materials such as ABS, PC, nylon-12 and PC-ABS blend may be affected by these factors, and they may have an effect on dimensional accuracy as well as surface finish. Therefore, determining the optimum values of temperature and humidity may need to be studied in various forms of materials.

There are some physical constraints imposed on FDM machine, which affect the selection of the optimal process settings and must be taken into consideration in the future research. The first constraint is that some FDM machines allow only four specific values of layer thickness which are 0.1270, 0.1778, 0.2540 and 0.3302 mm. Apart from these, the operator cannot select any other value because they are restricted by the nozzle diameter. The second practical constraint is that each nozzle diameter has its own raster width range.

The third practical constraint is that when the operator needs to use different number of contours, these are limited to the specific values. In this case, the operator will not be able to use any other values between this ranges. Therefore, optimization of FDM process parameters is complicated when many such constraints are also included. So, it will be difficult for the traditional DOEs to solve such kind of problem. Thus, developing new optimization approaches and mathematical modeling is needed to solve such type of constraints in order for the optimal parameters to be feasible and possible in practical applications [199].

The capability of the FDM to print composite filament allows it to be used for four-dimensional printing, rapid tooling, scaffolding, sensors and drug delivery device. The major issue faced during the fabrication of the microfluidic devices by FDM is the lack of resolution and transparency. Hence, studies towards a methodology for improving the resolution of FDM are demanded. Polymer products intended for medical application needs to be sterilized. Effect of sterilization on mechanical characteristics of FDM printed polymer products needs experimental investigation. Post-processing techniques have been proposed for improving the features of FDM parts, but most of the proposed approaches are material-specific, and there is a lack of a standard methodology that can process all types of material. Further, post-processing works can be extended for advanced materials like HIPS, PPSF, PEEK, composites, etc. The quantum of applications that are disclosed with the developed composite filament is limited, and it can be explored.

Most of the sensors are built from low melting point filaments like conductive PLA, ABS, TPU, etc., which cannot be used above 250°C. The possibility of using high-performance conductive composite like PEEK/CNT, PEEK/graphene, PEKK/graphene for sensing application can be studied. The sensing strength of a three-dimensional printed sensor is a function of filler type and its percentage of loading. The effect of filler concentration on different sensing mechanisms demands further research work. Brittleness of most of the conductive filaments renders its effective printing. Hence, investigation on the effect of adding different plasticizers on mechanical characteristics of three-dimensional printable filaments has to be explored.

FDM permits producing a scaffold of intricate design, but its osseointegration ability is less. Hence, the addition of reinforcement such as TiO2, TCP, PCL, HA, collagen is suggested. Further methods for achieving better functions from FDM printed scaffolds are demanded. Also, the inflammatory response of the fused deposition modelled scaffold has to be analysed in detail. The drug release rate of three-dimensional printed DDS is non-controllable. Suitable techniques for controlling the drug release rate of FDM made DDS are demanded. Besides, suitable parameters for developing and printing different drug-loaded filaments have to be identified.

The effort could be devoted to the development of filament for drug delivery devices, which can meet regulatory norms. The usage of FDM in microfluidics reduces the cost and enables the fabrication of complex profiles, but the presence of voids and high roughness affects the flow of fluids. Means of controlling these issues in microfluidics need to be addressed.

As most of the three-dimensional printer dimension is limited, various weld ability studies on FDM printed components are needed. The role of FDM in topology optimization demands further investigation. Modification of the printing head allows FDM to be used for the printing of continuous fibre-reinforced composites. However, the quantum of applications reported with FDM of the continuous fibre-reinforced composite is limited, and further research could be progressed in that direction as well [200].

Fabrication of lattice structures for tissue engineering with interconnecting pores for cell growth is possible by FDM. Further research is required in investigating printing characteristic of lattice structure manufactured via dynamic process parameter of the FDM process. Also, the biomechanical performance of these structures requires further evaluation.

Fabrication of dissimilar materials (plastic and metal) in single build needs further research as both materials have a different melting point. Combining them as per functional requirement can tackle strength and stiffness duality and solve many of the current problems incurred in different applications, for example, in automotive dashboard head impact analysis.

Since most of the engineering and high-performance FDM feedstocks are of the semicrystalline polymer type, there is a need to investigate the crystallization kinetics of these feedstocks during the cooling process as it also has a substantial influence on the dimensional accuracy, aesthetic quality as well as mechanical properties of printed parts. A controlled cooling rate is required to minimize the phenomenon of shrinkage and warping during the FDM process.

Table 10.1 summarizes the gaps identified in the literature review and presents a 13-phase roadmap for addressing existing shortcomings. These phases are broadly classified into three main areas, namely, pre-fabrication (Phases 1-4), fabrication (Phases 5-11) and post-fabrication (Phases 12-13).

Table 10.1: Road Map for Future Research [201]

Gaps related with	Phases	Research work to be undertaken
Design for manufacture	Phase 1	Guidelines for design for manufacture of FDM parts should be framed
Numerical modelling	Phase 2	Simulation to be performed using layer wise material properties that will aid in design of FDM parts.
	Phase 3	Rate of cooling and effect of environmental factors on FDM process should be numerically modelled so as to minimize approximation involved in FEA
Environmental factors	Phase 4	Optimal value of temperature and humidity for fabrication of FDM parts made up of various feedstock materials.
Machine	Phase 5	Air cavity minimization by adaptive feedback system by using sensor that modify tool path and process parameter values.
	Phase 6	Automatic prescription of process parameter values by using material and loading data so as to maximize material utilization.
	Phase 7	Forecast nozzle wear in case of feedstock material other than polymer so as to optimize use.
	Phase 8	Research work related to dynamic process parameter settings.
Material	Phase 9	Fabrication od dissimilar material in single build so as to optimize material uses in part as per desired response.
Process optimization	Phase 10	Process parameter optimization of various feedstock materials related to mechanical properties.
	Phase 11	Process parameter optimization for properties such as thermal, chemical, etc.
Non - destructive evaluation	Phase 12	In situ process monitoring for defect free component
PPFTs	Phase 13	Effect of metal coating and sterilization on parts should be studied.

10.2. Conclusion

Fused deposition modeling 3D printing is a versatile additive manufacturing technique that has emerged as one of the most prevalent techniques to have garnered the attention of the community. The FDM process has been extensively used by research and development community at universities and research organizations to further enhance the process and part quality, and to develop new materials for a variety of new engineering applications. Though the utilization of FDM has grown from prototype development to functional component fabrication, there are miles to go to achieve its full potential.

Usage of FDM in rapid tooling reduces the time and cost associated with traditional tooling fabrication, however, the strength of FDM based tooling is lower and as a result, the number of parts producible is lesser. It is observed that substantial work has been done in the optimization of process parameters for improving the mechanical and thermal properties of FDM parts. Efforts have also been made to improve the FDM process through numerical modelling. Research has also been done in FDM feedstock. FDM process has also been successfully used in fabricating implants for medical applications. Based on the literature review, direction for future work in FDM has been presented and roadmap for research is discussed. It is observed that much research work has been done on filament materials of ABS and PLA.

However, surface finish and mechanical properties of other materials should also be investigated. Effect of environmental factors on the quality and performance of FDM parts should also be further investigated. More realistic numerical simulation of build process should be modeled and validated. Complicated loading such as creep and fatigue needs further investigation to match real-world scenario. Fabrication of dissimilar material in the same build needs further research. Further, research efforts are also needed in technology advancement, new material development and automation of the FDM process for its wide acceptability in industries. Future research in FDM could increase the build velocity of the process, refine the texture of surfaces and advance of new materials, composites, plastics and biomaterials in filament shape for required industrial products. For reducing the manufacturing time of the process, future research could involve the combination of numerous extrusion nozzles or integrating logical build approaches where the interior of the part can be filled earlier using thicker rasters and external surfaces by finer rasters.

References

[1] https://www.totalmateria.com/page.aspx?ID=CheckArticle&site=ktn&NM=426

[2] https://www.ge.com/additive/additive-manufacturing

[3] https://www.industr.com/en/cost-efficiency-with-additive-manufacturing-2337955

[4] https://engineeringproductdesign.com/additive-manufacturing-process-steps/

[5] https://www.metal-am.com/introduction-to-metal-additive-manufacturing-and-3d-printing/

[6] https://www.wazp.io/2018/07/18/the-history-of-additive-manufacturing

[7] https://en.wikipedia.org/wiki/3D_printing

[8] Tanisha, John V Kennedy, Johan, (2019), "A comparison of traditional manufacturing vs additive manufacturing, the best method for the job". *Procedia Manufacturing,* Vol. 30, pp. 11-18.

[9] https://www.americanmachinist.com/shop-operations/article/21897894/ge-aviation-takes-on-additive-manufacturing.

[10] https://blog.animalventures.com/blog/additive-manufacturing-vs-traditionalmanufactu ring/#:~:text=Additive%20manufacturing%20operates%20by%20adding,contrast%2C%20are%20subtractive%20in%20nature.&text=Even%20though%20all%20additive%20manufacturing,different%20types%20of%20additive%20manufacturing.

[11] Ingrid Paoletti, (2017), "Mass customization with additive manufacturing: new perspectives for multi performative building components in architecture" *Procedia Engineering,* vol. 180, pp. 1150 – 1159.

[12] https://www.sculpteo.com/blog/2018/04/18/mass-customization-in-3d-printing-the-best-projects/

[13] https://www.k3syspro.com/advice-centre/erp-advice/whats-the-difference-between-3d -printing-and-additive-manufacturing/

[14] https://alloysilverstein.com/manufacturing-trends-additive-vs-traditional/

[15] https://insights.globalspec.com/article/7447/factors-to-consider-when-3d-printing-or-additive-manufacturing-metal-parts.

[16] Attaran, Mohsen. (2017). "The rise of 3-D printing: The advantages of additive manufacturing over traditional manufacturing". *Business Horizons* vol 60, Issue 5, September, pp. 677 - 688.

[17] https://www.cmtc.com/blog/benefits-of-additive manufacturing

[18] www.eltroquel.es/shop/impresion3d/impresion-3d-sla-1ml/

[19] https://www.3dsystems.com/on-demand-manufacturing/fast-turn-prototyping

[20] https://www.twi-global.com/technical-knowledge/faqs/what-is-3d-printing/pros-and-cons

[21] https://www.jabil.com/blog/design-for-additive manufacturing. html#:~:text=Design%20for%20Additive%20Manufacturing%20(DfAM,reducing%20weight%20and%20material%20consumption.

[22] https://www.theengineer.co.uk/additive-manufacturing-can-reduce/#:~:text=Additive%20manufacturing%20reduces%20waste,energy%20and%20create%20material%20waste.

[23] https://www.sculpteo.com/blog/2019/10/16/the-most-promising-3d-printed-organs-projects/

[24] https://www.metal-am.com/introduction-to-metal-additive-manufacturing-and-3d-printing/applications-for-additive-manufacturing-technology/

[25] https://www.3dnatives.com/en/applications-by-sector/#!

[26] https://www.cadcrowd.com/blog/how-3d-printed-prosthetics-are-changing-lives/

[27] S. L. N. Ford, (2014) "Additive Manufacturing Technology: Potential Implications for U S. Manufacturing," *Journal of International Commerce Econ.*, no. September, pp. 1–35.

[28] Y. Huang, M. C. Leu, J. Mazumder, and A. Donmez, "Additive Manufacturing: Current State, Future Potential, Gaps and Needs, and Recommendations," *Journal of Manufacturing Science Engineering*, vol. 137, no. 1, pp. 014001, 2015.

[29] https://www.3dhubs.com/knowledge-base/automotive-3d-printing-applications/

[30] B. Lyons, (2012) "Additive Manufacturing in Aerospace Examples and Research Outlook," *Front. Eng.*, vol. 42, no. 1, pp. 13–19.

[31] K. V. Wong and A. Hernandez, (2012) "A Review of Additive Manufacturing," ISRN Mech. Eng., vol. 2012, pp. 1–10.

[32] J. Coykendall, M. Cotteleer, L. Holdowsky, and M. Mahto, (2014) "3D opportunity in Aerospace and Defense".

[33] https://www.ge.com/additive/stories/new-manufacturing-milestone-30000-additive-fuel-nozzles

[34] Michael Siemer, (2013) "Adding to Your Tooling Options," Modern Machine Shop.

[35] https://www.eos.info/en/3d-printing-examples-applications/production-and-industry/tooling-diecast-molding

[36] Alexander Paolini, Stefan, Ernst, (2019), "Additive manufacturing in construction: A review on processes, applications, and digital planning methods" *Additive manufacturing,* vol. 30, pp. 100894.

[37] https://digital.hbs.edu/platform-rctom/wp-content/uploads/sites/4/2018/11/Pic-2-1.png

[38] http://home.iitk.ac.in/~jrkumar/download/Lecture-5.pdf

[39] https://www.lboro.ac.uk/research/amrg/about/the7categoriesofadditivemanufacturing/

[40] https://manufactur3dmag.com/7-types-additive-manufacturing-technologies/

[41] https://www.engineersgarage.com/tech-articles/3d-printing-processes-vat-photo-polymerisation-part-3-8/

[42] https://engineeringproductdesign.com/knowledge-base/vat-photo-polymerization/

[43] https://all3dp.com/2/main-types-additive-manufacturing/

[44] https://cdn.intechopen.com/pdfs/61889.pdf

[45] https://dailyengineers.com/stereolithography-3d-printing-technique/

[46] https://make.3dexperience.3ds.com/processes/material-extrusion

[47] Shah, J. & Snider, B. & Clarke, T. & Kozutsky, S. & Lacki, Maciej & Hosseini, Ali. (2019). Large-scale 3D printers for additive manufacturing: design considerations and challenges. *The International Journal of Advanced Manufacturing Technology,* 104. 4-6.

[48] Gonzalez-Gutierrez, J.; Cano, S.; Schuschnigg, S.; Kukla, C.; Sapkota, J.; Holzer, C. Additive Manufacturing of Metallic and Ceramic Components by the Material Extrusion of Highly-Filled Polymers: A Review and Future Perspectives. *Materials* 2018, 11, 840.

[49] https://omnexus.specialchem.com/selection-guide/3d-printing-and-additive-manufacturing-polymers-and-processes#MaterialExtrusion

[50] Milan Šljivić, et al., Additive manufacturing of functional parts based on material extrusion technology Contemporary Materials, VII–2 (2016).

[51] http://canadamakes.ca/what-is-powder-bed-fusion/

[52] https://engineeringproductdesign.com/knowledge-base/powder-bed-fusion/

[53] https://make.3dexperience.3ds.com/processes/powder-bed-fusion

[54] https://www.tctmagazine.com/additive-manufacturing-3d-printing-news/Stroke-patient-gets-life-back-with-3D-printed-cranial-implant-from-EOS/

[55] https://amfg.ai/2018/06/29/material-jetting-3d-printing-guide/

[56] http://canadamakes.ca/what-is-material-jetting/

[57] https://engineeringproductdesign.com/knowledge-base/material-jetting/

[58] Udroiu, Razvan & Braga, Cristian. (2017). Polyjet technology applications for rapid tooling. *MATEC Web of Conferences.* 112. 03011. 10.1051/matecconf/201711203011.

[59] https://www.additivemanufacturing.media/blog/post/am-101-nanoparticle-jetting-npj

[60] https://total3dprinting.org/category/guides-and-resources/

[61] https://www.treatstock.com/guide/article/126-what-is-material-jetting

[62] https://www.3dnatives.com/en/powder-binding100420174/#!

[63] https://make.3dexperience.3ds.com/processes/binder-jetting

[64] https://amfg.ai/2019/07/03/metal-binder-jetting-all-you-need-to-know/

[65] http://www8.hp.com/h20195/v2/GetPDF.aspx/4AA7-3333EEW.pdf

[66] https://amfg.ai/2018/03/13/3d-printing-binder-jetting-short-guide/

[67] https://www.additivemanufacturing.media/blog/post/hgan%C3%A4s-leverages-powder-expertise-into-binder-jetting-systems-now-for-sale

[68] https://www.twi-global.com/technical-knowledge/faqs/directed-energy-deposition

[69] https://www.digitalalloys.com/blog/directed-energy-deposition/

[70] https://engineeringproductdesign.com/knowledge-base/direct-energy-deposition/

[71] https://optomec.com/how-3d-metal-printing-saves-time-and-lowers-costs-ded-for-repair-of-industrial-components/

[72] https://www.3dprinting.lighting/3d-printing-technologies/sheet-lamination/

[73] https://engineeringproductdesign.com/knowledge-base/sheet-lamination/#:~:text=Mat erial%20suitability,to%20activate%20the%20adhesive%20layer.

[74] https://www.3dnatives.com/en/fused-deposition-modeling100420174/

[75] https://www.marketsandmarkets.com/Market-Reports/3d-printing-market-1276.html

[76] https://www.livescience.com/39810-fused-deposition-modeling.html

[77] https://www.iitk.ac.in/dordold/index.php?option=com_content&view=category&layout=blog&id=200&Itemid=219

[78] Tran, N., Van-Nam Nguyen, A. Ngo and Van-Cuong Nguyen. (2017) "Study on the Effect of Fused Deposition Modeling (FDM) Process Parameters on the Printed Part Quality."

[79] https://pick3dprinter.com/3d-printer-parts/

[80] https://www.creativitybuzz.org/fused-deposition-modeling/

[81] https://3dinsider.com/3d-printer-extruders/#:~:text=The%20extruder%20hotend%20is

%20one,to%20form%20a%20solid%20object.&text=Size%2Dwise%2C%20the%20hotend%20is,print%20quality%20and%20material%20compatibility.

[82] https://www.amazon.com/DisTech-Automation-Prometheus-Hot-End/dp/B01GUEBB B6/ref=as_li_ss_tl?ie=UTF8&linkCode=ll1&tag=insider0420&linkId=5de3006cbac77ca7c0f3382f59a02608&language=en_US

[83] https://blog.prusaprinters.org/prusa-research-summer-update-2018_8535/

[84] https://www.reprap.me/15-20-25mm-10mm-bore-fdm-aluminum-heating-block-for-extruder-mk10-nozzle.html

[85] https://www.amazon.com/Printer-0-4mm-Extruder-Nozzle-175mm/dp/B00L4UXQ42

[86] https://filament2print.com/gb/blog/36_types-3d-extruders-and-hotend.html#:~:text=W ithin%20extruders%20there%20are%20two,as%20in%20the%20Titan%20Aero.

[87] https://www.mdpi.com/1999-4923/10/4/203/htm

[88] https://all3dp.com/1/3d-printer-extruder-nozzle-guide/

[89] https://www.3dnatives.com/en/four-types-fdm-3d-printers140620174/#!

[90] https://www.fargo3dprinting.com/choose-cartesian-delta-3d-printers/

[91] https://3dprint.com/219929/printrbot-closes-doors/

[92] https://manufactur3dmag.com/understanding-the-four-types-of-fdm-3d-printers-cartesian-delta-polar-scara/

[93] https://3dfactorychina.en.made-in-china.com/product/KCumaRPLgFhp/China-3D-Metal-Printer-Power-off-Resume-Print-Fully-Assembled-Mini-Desktop-Delta-DIY-Fdm-3D-Printer.html

[94] https://3dinsider.com/polar-3d-printers/#:~:text=A%20polar%203D%20printer%20lo oks,most%20notable%20contrast%20in%20design.&text=In%20other%20polar%203D%20printer,rotating%20in%20a%20stationary%20position.

[95] https://us.vwr.com/store/product/21104232/polar-3d-printer-and-polar-cloud

[96] https://3dprint.com/85215/branch-3d-printed-walls/

[97] https://amfg.ai/2018/08/15/3d-printing-and-robotics-create-smart-factories/

[98] https://3dinsider.com/3d-printing-bed/

[99] https://apm-designs.com/3d-printer-main-components/

[100] https://www.mdpi.com/2227-7080/5/2/36/htm

[101] http://www.materialseducation.org/educators/matedu-modules/docs/Materials_in_FDM.pdf

[102] http://www.3dprinterpro.com/

[103] https://www.3dhubs.com/knowledge-base/fdm-3d-printing-materials-compared/

[104] https://diyconnect.net/popular-filaments-used-in-fdm-3d-printing-technology/

[105] https://www.3dnatives.com/en/pla-3d-printing-guide-190820194/#!

[106] https://www.simplify3d.com/support/materials-guide/pla/

[107] https://www.sculpteo.com/en/glossary/absdefinition/#:~:text=Acrylonitrile%20Butadiene%20Styrene%20 (ABS)%20is,in%20order%20to%20liquefy%20it.

[108] https://www.xometry.com/capabilities/fused-deposition-modeling/

[109] https://www.materialise.com/en/manufacturing/materials/polycarbonate

[110] https://www.beamler.com/3d-printing-capabilities/materials/pc-polycarbonate/

[111] https://www.javelin-tech.com/3d/stratasys-materials/pc/

[112] https://vexmatech.com/PC-polycarbonate-FDM-technology-3dprinting-material.html

[113] https://all3dp.com/1/3d-printing-materials-guide-3d-printer-material/

[114] Gajdos, Ivan & Maňková, Ildikó & Jachowicz, Tomasz & Tor-Świątek, Aneta. (2016). Application of Rapid Tooling approach in process of thermoforming mold production. *International Engineering Symposium at Bánki.* 8.

[115] https://www.makerbot.com/stories/engineering/everything-you-need-to-know-about-petg-3d-printing/

[116] https://amfg.ai/2019/02/12/fdm-3d-printing-asa-petg-pc-filaments-compared/#:~:text =PETG%20is%20a%20variant%20of,properties%20of%20ABS%20and%20PLA

[117] https://www.treatstock.com/material/petg

[118] https://all3dp.com/2/tpu-filament-explained-and-compared/

[119] https://www.stratasysdirect.com/fr/technologies/fused-deposition-modeling/fdm-tpu-92a-elastomer

[120] https://tractus3d.com/materials/tpu

[121] https://filament2print.com/gb/blog/43_asa-3d-printing.html

[122] https://investors.stratasys.com/news-events/press-releases/detail/232/new-stratasys-asa-thermoplastic-is-uv-resistant-and-has

[123] https://tractus3d.com/materials/ultem

[124] https://3dprintingservices.co.nz/product/pei-ultem-9085-3d-filament-nz/

[125] https://www.3dnatives.com/en/ultem-030820204/#!

[126] S Wickramasinghe, T Do, P Tran, "FDM-based 3D printing of polymer and associated composite: A review on mechanical properties, defects and treatments", *Polymers*, 2020.

[127] Onwubolu G. C. and Rayegani F. "Characterization and optimization of mechanical properties of ABS parts manufactured by the fused deposition modelling process", *International Journal of Manufacturing Engineering*, 1-13, (2014).

[128] Rayegani, Farzad & Onwubolu, Godfrey. (2014). Fused deposition modelling (FDM) process parameter prediction and optimization using group method for data handling (GMDH) and differential evolution (DE). *The International Journal of Advanced Manufacturing Technology*. 73. 509-519. 10.1007/s00170-014-5835-2.

[129] Alsoufi, Mohammad & El-Sayed, Abdulrhman. (2018). Surface Roughness Quality and Dimensional Accuracy - A Comprehensive Analysis of 100% Infill Printed Parts Fabricated by a Personal/Desktop Cost-Effective FDM 3D Printer. *Materials Sciences and Applications*. 9. 11-40. 10.4236/msa.2018.91002.

[130] Campbell, R.I.; Martorelli, M.; Lee, H.S. Surface roughness visualisation for rapid prototyping models. *Comput. Aided Des. Appl.*, 2002, 34, 717–725.

[131] https://www.fictiv.com/articles/ultimate-guide-to-finishing-3d-printed-parts

[132] Sung-Hoon Ahn, Michael Montero, Dan Odell, Shad Roundy, Paul K. Wright, (2002) "Anisotropic material properties of fused deposition modeling ABS", *Rapid Prototyping Journal*, Vol. 8 Issue: 4, pp. 248-257.

[133] Daiyan, H. et al., 2012. Shear Testing of Polypropylene Materials Analysed by Digital Image Correlation and Numerical Simulations. *Experimental Mechanics*, 52(9), pp. 1355–1369.

[134] Jason Cantrell1, Sean Rohde1, David Damiani2, Rishi Gurnani3, Luke DiSandro1" *Experimental Characterization of the Mechanical Properties of 3D-Printed ABS and Polycarbonate Parts.*

[135] Divyathej M V, Varun M, Rajeev P, Analysis of mechanical behavior of 3D printed ABS parts by experiments, *International Journal of Scientific & Engineering Research*, Volume 7, Issue 3, March-2016

[136] https://www.sculpteo.com/en/glossary/pla-definition/

[137] Esposito Corcione, C.; Gervaso, F.; Scalera, F.; Montagna, F.; Sannino, A.; Maffezzoli, A. The feasibility of printing polylactic acid–nanohydroxyapatite composites using a low-cost fused deposition modeling 3D printer. *J. Appl. Polym. Sci.*, 2017, 134.

[138] Zhuang, Y.; Song, W.; Ning, G.; Sun, X.; Sun, Z.; Xu, G.; Zhang, B.; Chen, Y.; Tao, S. 3D-printing of materials with anisotropic heat distribution using conductive polylactic acid composites. *Mater. Des.*, 2017, 126,135–140.

[139] Chen, Q.; Mangadlao, J.D.; Wallat, J.; De Leon, A.; Pokorski, J.K.; Advincula, R.C. 3D printing biocompatible polyurethane/poly (lacticacid)/graphene oxide nanocomposites: Anisotropic properties. *ACS Appl. Mater. Interfaces* 2017, 9, 4015–4023.

[140] Prashantha, K.; Roger, F. Multifunctional properties of 3D printed poly (lactic acid)/graphene nanocomposites by fused deposition modeling. *J. Macromol. Sci. A* 2017, 54, 24–29.

[141] Tian, X.; Liu, T.; Wang, Q.; Dilmurat, A.; Li, D.; Ziegmann, G. Recycling and remanufacturing of 3D printed continuous carbon fiber reinforced PLA composites. *J Clean. Prod.*, 2017, 142, 1609–1618.

[142] Cicala, G.; Giordano, D.; Tosto, C.; Filippone, G.; Recca, A.; Blanco, I. Polylactide (PLA) Filaments a Biobased Solution for Additive Manufacturing: Correlating Rheology and Thermomechanical Properties with Printing Quality. *Materials* 2018, 11, 1191.

[143] Altan, M.; Eryildiz, M.; Gumus, B.; Kahraman, Y. Effects of process parameters on the quality of PLA products fabricated by fused deposition modeling (FDM): Surface roughness and tensile strength. *Mater. Test.*, 2018, 60, 471–477.

[144] Carou, Diego & Pérez, Mercedes & Medina-Sánchez, G. & García-Collado, Alberto & Gupta, Munish. (2018). Surface Quality Enhancement of Fused Deposition Modeling (FDM) Printed Samples Based on the Selection of Critical Printing Parameters. *Materials.* 11.

[145] Cristina Vălean, Liviu Marşavina, Mihai Mărghitaş, Emanoil Linul, Javad Razavi, Filippo Berto, Effect of manufacturing parameters on tensile properties of FDM printed specimens, *Procedia Structural Integrity,* Volume 26, 2020, Pages 313-320.

[146] Brischetto, Salvatore & Torre, Roberto. (2020). Tensile and Compressive Behavior in the Experimental Tests for PLA Specimens Produced via Fused Deposition Modelling Technique. *Journal of Composites Science.* 4. 140. 10.3390/jcs4030140.

[147] Padovano, Elisa & Galfione, Marco & Concialdi, Paolo & Lucco, Gianni & Badini, C. (2020). Mechanical and Thermal Behavior of Ultem® 9085 Fabricated by Fused-Deposition Modeling. *Applied Sciences.* 10. 3170. 10.3390/app10093170.

[148] Byberg, K.I.; Gebisa, A.W.; Lemu, H.G. Mechanical properties of ULTEM 9085 material processed by fused deposition modeling. *Polym. Test.*, 2018, 72, 335–347.

[149] Zaldivar, R.J.; Mclouth, T.D.; Ferrelli, G.L.; Patel, D.N.; Hopkins, A.R.; Witkin, D. Effect of initial filament moisture content on the microstructure and mechanical performance of ULTEM® 9085 3D printed parts. *Addit. Manuf.*, 2018, 24, 457–466.

[150] Zaldivar, R.J.; Witkin, D.B.; Mclouth, T.; Patel, D.N.; Schmitt, K.; Nokes, J.P. Influence of processing and orientation print effects on the mechanical and thermal behavior of 3D-Printed ULTEM 9085 Material. *Addit. Manuf.*, 2017, 13, 71–80.

[151] Motaparti, Krishna Prasanth & Taylor, Gregory & Leu, Ming & Chandrashekhara, K. (2015). Effects of build parameters on compression properties for ULTEM 9085 parts by fused deposition modeling.

[152] Byberg, Kate & Gebisa, Aboma & Lemu, Hirpa. (2018). Mechanical properties of ULTEM 9085 material processed by fused deposition modeling. Polymer Testing. 72.

[153] A. Bagsik, V. Schöppner, Mechanical properties of fused deposition modeling parts manufactured with Ultem*9085, *Proceedings of the Society of Plastics Engineers ANTEC Conference* (Anaheim), Boston, MA, USA, 8–10 May, 2011, pp. 1–5.

[154] Brenken B, Barocio E, Favaloro A, Kunc V, Pipes RB. Fused filament fabrication of fiber- reinforced polymers: a review. *Add Manuf.*, 2018; 21:1–16.

[155] Rahim T, Abdullah AM, Akil HM. Recent developments in fused deposition modelling based 3D printing of polymers and their composites. *Polym Rev.*, 2019; 59: 589–624.

[156] Heller B, Smith DE, Jack DA. Effects of extrudate swell and nozzle geometry on fiber orientation in fused filament fabrication nozzle flow. *Add Manuf.*, 2016; 12: 252–64.

[157] Kabir SMF, Mathur K, Seyam AFM. A critical review on 3D printed continuous fiber-reinforced composites: history, mechanism, materials and properties. *Compos Struct.*, 2020; 232: 111476.

[158] Ahn S, Montero M, Odell D, Roundy S, Wright P. Anisotropic material properties of fused deposition modeling ABS. *Rapid Prototyp J.*, 2002; 8: 248–57.

[159] Dawoud M, Taha I, Ebeid SJ. Mechanical behaviour of ABS: an experimental study using FDM and injection moulding techniques. *J Manuf Process.*, 2016; 21: 39–45.

[160] Zhong W, Li F, Zhang Z, Song L, Li Z. Short fiber reinforced composites for fused deposition modeling. *Mater Sci Eng.*, A 2001; 301: 125–30.

[161] Shofner M, Rodrí Guez-Mací as F, Vaidyanathan R, Barrera E. Single wall nanotube and vapour grown carbon fiber reinforced polymers processed by extrusion freeform fabrication. *Composites Part A Appl Sci Manuf.*, 2003; 34: 1207–17.

[162] Shofner ML, Lozano K, Rodríguez-Macías FJ, Barrera EV. Nanofiber-reinforced polymers prepared by fused deposition modeling. *J Appl Polym Sci.*, 2003; 89: 3081–90.

[163] Ning F, Cong W, Hu Y, Wang H. Additive manufacturing of carbon fiber-reinforced plastic composites using fused deposition modeling: effects of process parameters on tensile properties. *J Compos Mater* 2016; 51: 451–62.

[164] Tekinalp HL, Kunc V, Velez-Garcia GM, Duty CE, Love LJ, Naskar AK, et al. Highly oriented carbon fiber–polymer composites via additive manufacturing. *Compos Sci Technol.*, 2014; (105): 144–50.

[165] Zhang W, Cotton C, Sun J, Heider D, Gu B, Sun B, et al. Interfacial bonding strength of short carbon fiber/acrylonitrile-butadiene-styrene composites fabricated by fused deposition modeling. *Compos B Eng.*, 2017; 2018(137): 51–9.

[166] Love LJ, Kunc V, Rios O, Duty CE, Elliott AM, Post BK, Blue CA. The importance of carbon fiber to polymer additive manufacturing. *J Mater Res.*, 2017; 29: 1893–8.

[167] Weng Z, Wang J, Senthil T, Wu L. Mechanical and thermal properties of ABS/montmorillonite nanocomposites for fused deposition modeling 3D printing. *Mater Des.*, 2016; 102: 276–83.

[168] Wei X, Li D, Jiang W, Gu Z, Wang X, Zhang Z, et al. 3D printable graphene composite. *Sci Rep.*, 2015; 5: 11181.

[169] Dul S, Fambri L, Pegoretti A. Fused deposition modelling with ABS-graphene nano composites. *Composites Part A* 2016; 85: 181–91.

[170] Aumnate C, Pongwisuthiruchte A, Pattanauwat P, Potiyaraj P. Fabrication of ABS/graphene oxide composite filament for fused filament fabrication 3D printing. *Adv. Mater. Sci. Engg.*, 2018: 2830437.

[171] Plymill A, Minneci R, Alexander Greeley D, Gritton J, Alexander D, Greeley D. *Graphene PLA composite feedstock development for fused deposition Modeling.* Thesis, univ. Tennessee. 2016.

[172] Bustillos J, Montero D, Nautiyal P, Loganathan A, Boesl B, Agarwal A. Integration of graphene in poly(lactic) acid by 3D printing to develop creep and wear-resistant hierarchical nanocomposites. *Polym Compos.*, 2018; 39: 3877–88.

[173] Sweeney CB, Lackey BA, Teipel BR, Saẹd MA, Moran AG, Achee TC, et al. Welding of 3D-printed carbon nanotube–polymer composites by locally induced microwave heating. *Sci Asia* 2017; 3: e1700262.

[174] Ivanov E, Kotsilkova R, Xia H, Chen Y, Donato RK, Donato K, et al. PLA/Graphene/MWCNT composites with improved electrical and thermal properties suitable for FDM 3D printing applications. *Appl Sci.*, 2019; 9: 1209.

[175] Li N, Li Y, Liu S. Rapid prototyping of continuous carbon fiber reinforced polylactic acid composites by 3D printing. *J Mater Process Technol.*, 2016; 238: 218–25.

[176] Tian X, Liu T, Yang C, Wang Q, Li D. Interface and performance of 3D printed continuous carbon fiber reinforced PLA composites. *Composites Part A Appl Sci Manuf.*, 2016; (88): 198–205.

[177] Tian X, Liu T, Wang Q, Dilmurat A, Li D, Ziegmann G. Recycling and remanufacturing of 3D printed continuous carbon fiber reinforced PLA composites. *J Clean Prod.*, 2017; 142:1609–18.

[178] Singh R, Singh S, Fraternali F. Development of in-house composite wire based feed stock filaments of fused deposition modelling for wear-resistant materials and structures. *Compos B Eng.*, 2016; (98): 244–9.

[179] Abdullah AM, Tuan Rahim TNA, Mohamad D, Akil HM, Rajion ZA. Mechanical and physical properties of highly ZrO2/β-TCP filled polyamide 12 prepared via fused deposition modelling (FDM) 3D printer for potential craniofacial reconstruction application. *Mater Lett.*, 2017; 189: 307–9.

[180] Zhu D, Ren Y, Liao G, Jiang S, Liu F, Guo J, et al. Thermal and mechanical properties of polyamide 12/graphene nanoplatelets nanocomposites and parts fabricated by fused deposition modeling. *J Appl Polym Sci.*, 2017; 134: 45332.

[181] Goh GD, Dikshit V, Nagalingam AP, Goh GL, Agarwala S, Sing SL, et al. Characterization of mechanical properties and fracture mode of additively manufactured carbon fiber and glass fiber reinforced thermoplastics. *Mater Des.*, 2018; (137): 79–89.

[182] Dickson AN, Barry JN, McDonnell KA, Dowling DP. Fabrication of continuous carbon, glass and Kevlar fibre reinforced polymer composites using additive manufacturing. *Addit Manuf.*, 2017; (16):146–52.

[183] Carneiro OS, Silva AF, Gomes R. Fused deposition modeling with polypropylene. *Mater Des.*, 2015; 83:768–76.

[184] Spoerk M, Sapkota J, Weingrill G, Fischinger T, Arbeiter F, Holzer C. Shrinkage and warpage optimization of expanded perlite filled polypropylene composites in extrusion based additive manufacturing. *Macromol Mater Eng.*, 2017; 10:1700143.

[185] Stoof D, Pickering K. Sustainable composite fused deposition modelling filament using recycled pre-consumer polypropylene. *Compos B Eng.*, 2018; 135:110–8.

[186] Schirmeister CG, Hees T, Licht EH, Mulhaupt R. 3D printing of high density Polyethylene by filament fabrication. *Add Manuf.*, 2019; 28:152–9.

[187] Bedi P, Singh R, Ahuja IPS. Investigations for tool life of 3D printed HDPE and LDPE composite based rapid tooling for thermoplastics machining applications. *Engg Res Express* 2019: 1. 015003.

[188] Tarres Q, Melbo JK, Delgado-Aguilar M, Espinach FX, Mutje P, Chinga- Carrasco G. Bio - polyethylene reinforced with thermomechanical pulp fibers: Mechanical and micromechanical characterization and its application in 3D-printing by fused deposition modeling 2018; 153: 70–7.

[189] Kishore V, Chen X, Ajinjeru C, Hassen AA, Lindal J, Failla J, et al. Additive Manufacturing of high performance semi crystalline thermoplastic and their composites, *Proceedings of 27th annual international solid freeform fabrication symposium.* 2017.

[190] Beretta S, Davies R, Shyng YT, Wang Y. Ghita O, Fused deposition modelling of high temperature polymers: exploring CNT PEEK composites. *Polym Test.*, 2017; 63:251–62.

[191] https://www.3deo.co/featured/metal3dprintingprocessesmetalextrusionfffdm/#:~:text=Applications%20of%20Metal%20AM%20Extrusion&text=This%20is%20an%20extremely%20capable,for%20cost%2Deffective%20metal%20prototyping.

[192] https://www.google.com/imgres?imgurl=https%3A%2F%2Fwww.createitreal.com%2Fwpcontent%2Fuploads%2F2018%2F07%2Ffdm00x514.jpg&imgrefurl=https%3A%2F%2Fwww.createitreal.com%2F2018%2F07%2F13%2Frealvision-slicer-getting-ready-for-metalfdm3dprinting%2F&tbnid=glsi9_3T8aCt3M&vet=12ahUKEwiM2W38tXuAhUaIisKHfj3BqoQMygGegUIARCwAQ..i&docid=saHe3DCClrrIRM&w=900&h=514&q=metal%20fdm%20parts&ved=2ahUKEwiM2W38tXuAhUaIisKHfj3BqoQMygGegUIARCwA

[193] https://www.createitreal.com/2018/07/13/realvision-slicer-getting-ready-for-metal-fdm-3d-printing/

[194] https://www.3dhubs.com/guides/metal-3d-printing/#metal-extrusion

[195] Liu, Bin & Wang, Yuxiang & Lin, Ziwei & Zhang, Tao. (2019). Creating metal parts by Fused Deposition Modeling and Sintering. *Materials Letters.* 263. 127252.

[196] S.H. Masood and W.Q. Song, Development of new metal/ polymer materials for rapid tooling using fused deposition modelling, *Material and Design* 25 2004.

[197] Hwang, Seyeon & Reyes, Edgar & Moon, Kyoung-sik & Rumpf, Raymond & Kim, Nam. (2015). Thermo-mechanical Characterization of Metal/Polymer Composite Filaments and Printing Parameter Study for Fused Deposition Modeling in the 3D Printing Process. *Journal of Electronic Materials.* 44.

[198] Nabipour, Mojtaba & Akhoundi, Behnam & Bagheri, Arvin. (2019). Manufacturing of polymer/metal composites by fused deposition modeling process with polyethylene. *Journal of Applied Polymer Science.*

[199] Mohamed, Omar. (2015). Optimization of fused deposition modeling process parameters: a review of current research and future prospects. *Advances in Manufacturing.* 3. 42-52.

[200] S, Anoop & Ponnusamy, Senthil & T., Sathies. (2020). A review on advancements in applications of fused deposition modelling process. 10.1108/RPJ-08-2018-0199.

[201] Vyavahare, Swapnil & Teraiya, Soham & Panghal, Deepak & Kumar, Shailendra. (2019). Fused deposition modelling: a review. *Rapid Prototyping Journal.*